Allen Hazen

The Filtration of Public Water-Supplies

Allen Hazen

The Filtration of Public Water-Supplies

ISBN/EAN: 9783337141035

Printed in Europe, USA, Canada, Australia, Japan

Cover: Foto ©berggeist007 / pixelio.de

More available books at **www.hansebooks.com**

THE FILTRATION

OF

PUBLIC WATER-SUPPLIES.

BY

ALLEN HAZEN,

LATE CHEMIST IN CHARGE OF THE LAWRENCE EXPERIMENT STATION OF THE MASSACHUSETTS
STATE BOARD OF HEALTH, AND CHEMIST OF THE DEPARTMENT OF WATER-SUPPLY
AND SEWERAGE OF THE WORLD'S COLUMBIAN EXPOSITION; MEMBER OF
THE BOSTON SOCIETY OF CIVIL ENGINEERS, THE NEW ENGLAND
WATER-WORKS ASSOCIATION, THE AMERICAN PUBLIC
HEALTH ASSOCIATION, ETC.

SECOND EDITION.
FIRST THOUSAND.

NEW YORK:
JOHN WILEY & SONS.
LONDON: CHAPMAN & HALL, LIMITED.
1896.

ROBERT DRUMMOND, ELECTROTYPER AND PRINTER, NEW YORK.

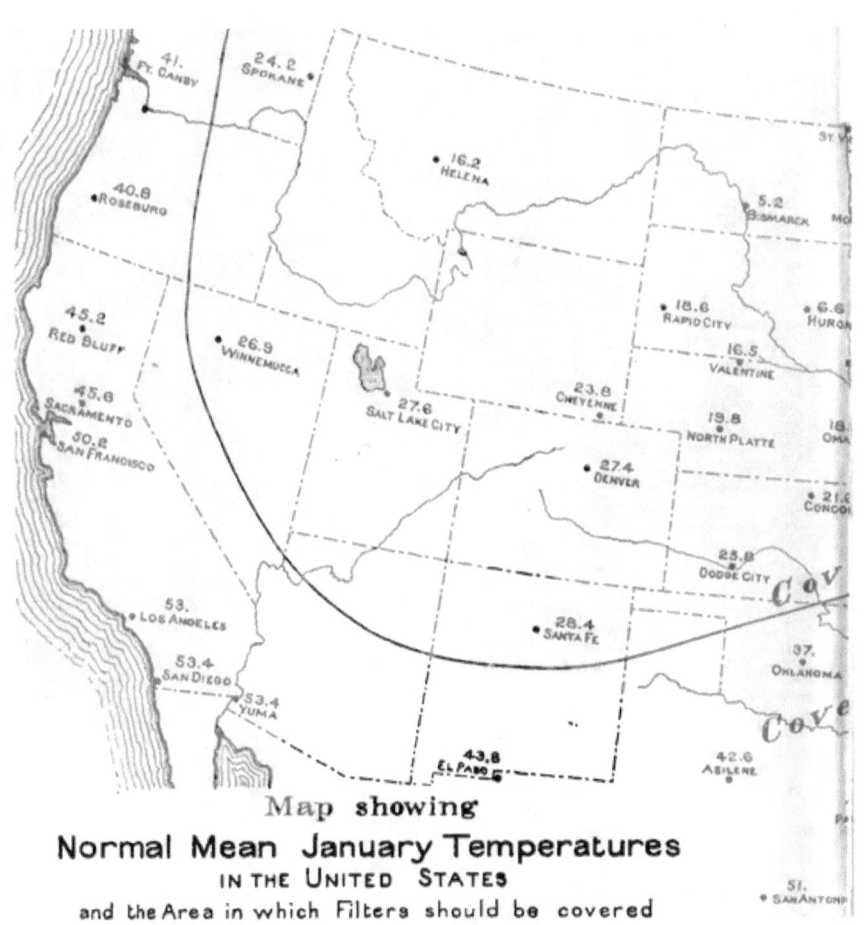

Map showing
Normal Mean January Temperatures
IN THE UNITED STATES
and the Area in which Filters should be covered

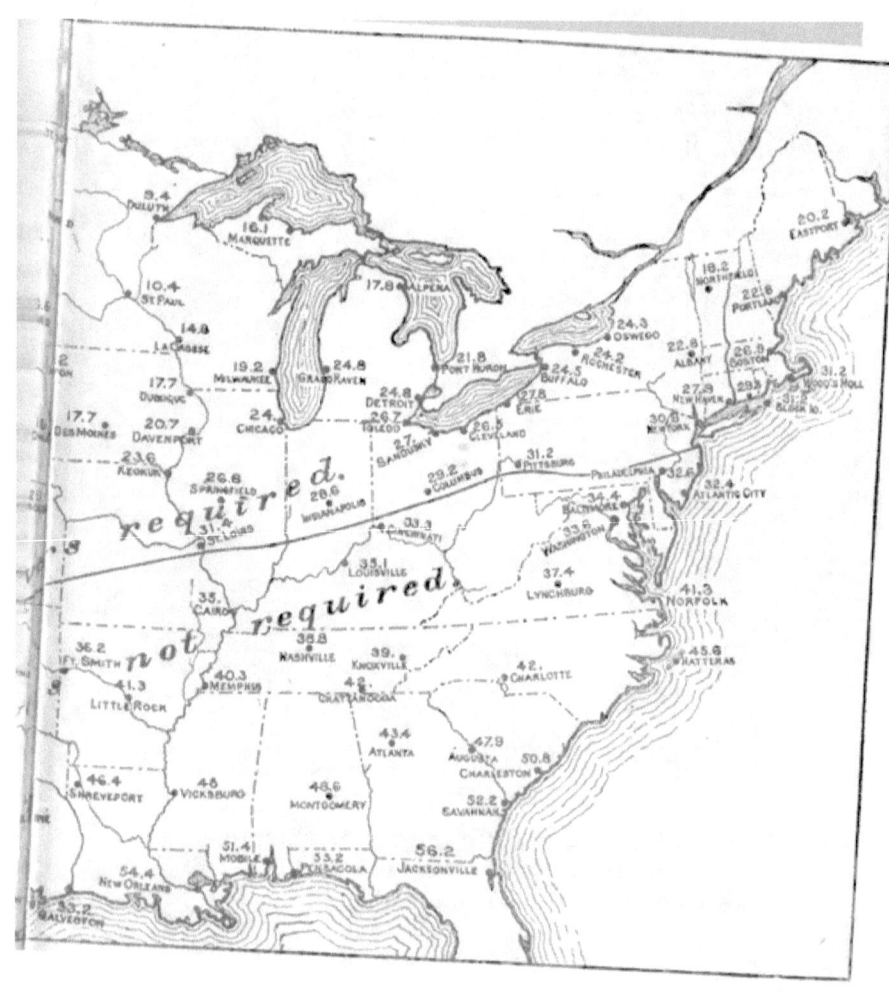

PREFACE.

THE subject of water-filtration is commencing to receive a great deal of attention in the United States. The more densely populated European countries were forced to adopt filtration many years ago, to prevent the evils arising from the unavoidable contaminations of the rivers and lakes which were the only available sources for their public water-supplies; and it has been found to answer its purpose so well that at the present time cities in Europe nearly if not quite equal in population to all the cities of the United States are supplied with filtered water.

Many years ago, when the whole subject of water-supply was still comparatively new in this country, filtration was considered as a means for rendering the waters of our rivers suitable for the purpose of domestic water-supply. St. Louis investigated this subject in 1866, and the engineer of the St. Louis Water Board, the late Mr. J. P. Kirkwood, made an investigation and report upon European methods of filtration which was published in 1869, and was such a model of full and accurate statement combined with clearly-drawn conclusions that, up to the present time, it has remained the only treatise upon the subject in English, notwithstanding the great advances which have been made, particularly in the last ten years, with the aid of knowledge of the bacteria and the germs of certain diseases in water.

Unfortunately the interest in the subject was not maintained in America, but was allowed to lag for many years; it was cheaper to use the water in its raw state than it was to purify it; the people became indifferent to the danger of such use, and the disastrous epidemics of cholera and typhoid fever, as well as of minor diseases, which so often resulted from the use of polluted water, were attributed to other causes. With increasing study and diffusion of knowledge the relations of water and disease are becoming better known, and the present state of things will not be allowed to continue; indeed at present there is inquiry at every hand as to the methods of improving waters.

The one unfortunate feature is the question of cost. Not that the **cost of filtration is** excessive or beyond the means of American **communities; in point of** fact, exactly the reverse is the case; **but we have been so** long accustomed to obtain drinking-water without expense other than pumping that any cost tending to improved quality seems excessive, thus affording a chance for the installation of inferior filters, which by failing to produce the promised results tend to bring the whole process into disrepute, since few people can distinguish between an adequate filtration and a poor substitute for it. It is undoubtedly true that improvements are made, and will continue to be made, in processes of filtration; so it will often be possible to reduce the expense of the process without decreasing the efficiency, but great care must be exercised in such cases to maintain the conditions really essential to success.

In the present volume I have endeavored to explain briefly the nature of filtration and the conditions which, in half a century of European practice, have been found essential for successful practice, with a view of stimulating interest in the subject, and of preventing the unfortunate and disappointing results which so easily result from the construction of inferior filters. The economies which may possibly result by the use of an inferior filtration are comparatively small, and it is believed that in those American cities where filtration is necessary or desirable it will be found best in every case to furnish **filters of** the best construction, fully able to do what is required **of them** with ease and certainty.

CONTENTS.

	PAGE
CHAPTER I. INTRODUCTION	1
II. CONTINUOUS FILTERS AND THEIR CONSTRUCTION	5
Sedimentation-basins	8
Size of Filter-beds	10
Covers for Filters	12
III. FILTERING-MATERIALS	19
Sand	19
Gravel	31
Underdrains	35
Depth of Water on Filters	41
IV. RATE OF FILTRATION AND LOSS OF HEAD	43
Rate of Filtration	43
Loss of Head and Apparatus for regulating it	48
Limit to the Loss of Head	56
V. CLEANING FILTERS	64
Scraping	64
Frequency of Scraping	68
Sand-washing	72
VI. THEORY AND EFFICIENCY OF FILTRATION	79
Bacterial Examination of Waters	89
VII. INTERMITTENT FILTRATION	93
The Lawrence Filter	96
The Chemnitz Filter	100
VIII. OTHER METHODS OF FILTRATION	106
Mechanical Filters without Coagulents	106
The Use of Alum	109
Precipitated Alumina and other Chemicals	113
The Use of Metallic Iron	114
Household Filters	115
IX. COST AND ADVANTAGES OF FILTRATION	118
Cost of Filtration	118
Objects of Filtration	122
What Waters require Filtration?	130
X. CONCLUSIONS	133
Water and Disease	133

		PAGE
APPENDIX I. GERMAN OFFICIAL REGULATION IN REGARD TO FILTRATION		139
II. EXTRACTS FROM DR. REINCKE'S REPORT UPON THE HEALTH OF HAMBURG FOR 1892		144
III. METHODS OF SAND-ANALYSIS		151
IV. STATISTICS OF SOME FILTERS		159
V. WATER-SUPPLY OF LONDON		161
VI. WATER-SUPPLY OF BERLIN		167
VII. WATER-SUPPLY OF ALTONA		171
VIII. WATER-SUPPLY OF HAMBURG		175
IX. NOTES ON SOME OTHER EUROPEAN SUPPLIES		178
X. LITERATURE OF FILTRATION		183
INDEX		191

UNITS EMPLOYED.

The units used in this work are uniformly those in common use in America, with the single exception of data in regard to sand-grain sizes, which are given in millimeters. The American units were not selected because the author prefers them or considers them particularly well suited to filtration, but because he feared that the use of the more convenient metric units in which the very comprehensive records of Continental filter plants are kept would add to the difficulty of a clear comprehension of the subject by those not familiar with those units, and so in a measure defeat the object of the book.

TABLE OF EQUIVALENTS.

Unit.	Metric Equivalent.	Reciprocal.
Foot	0.3048 meter	3.2808
Mile	1609.34 meters	0.0006214
Acre	4047 square meters	0.0002471
Gallon*	3.785 liters	0.26417
1 million gallons	3785 cubic meters	0.00026417
Cubic yard	0.7645 cubic meters	1.308
1 million gallons per acre daily	0.9354 { meter in depth of water daily }	1.070

* The American gallon is 231 cubic inches or 0.8333 of the imperial gallon. In this work American gallons are always used, and English quantities are stated in American, not imperial, gallons.

ACKNOWLEDGMENT.

I WISH to acknowledge my deep obligation to the large number of European engineers, directors, and superintendents of water-works, and to the health officers, chemists, bacteriologists, and other officials who have kindly aided me in studying the filtration-works in their respective cities, and who have repeatedly furnished me with valuable information, statistics, plans, and reports.

To mention all of them would be impossible, but I wish particularly to mention Major-General Scott, Water-examiner of London; Mr. Mansergh, Member of the Royal Commission on the Water-supply of the Metropolis; Mr. Bryan, Engineer of the East London Water Company; and Mr. Wilson, Manager of the Middlesborough Water-works, who have favored me with much valuable information.

In Holland and Belgium I am under special obligations to Messrs. Van Hasselt and Kemna, Directors of the water companies at Amsterdam and Antwerp respectively; to Director Stang of the Hague Water-works; to Dr. Van't Hoff, Superintendent of the Rotterdam filters; and to my friend H. P. N. Halbertsma, who, as consulting engineer, has built many of the Dutch water-works.

In Germany I must mention Profs. Frühling, at Dresden, and Flügge, at Breslau; Andreas Meyer, City Engineer of Hamburg; and the Directors of water-works, Beer at Berlin, Dieckmann at Magdeburg, Nau at Chemnitz, and Jockmann at Liegnitz, as well as the Superintendent Engineers Schroeder at Hamburg, Debusmann at Breslau, and Anklamm and Piefke at Berlin, the latter the distinguished head of the Stralau works, the first and most widely known upon the Continent of Europe.

I have to acknowledge my obligation to City Engineer Sechner at Budapest, and to the Assistant Engineer in charge of water-works, Kajlinger; to City Engineer Peters and City Chemist Bertschinger at Zürich; and to Assistant Engineer Regnard of the Compagnie Générale des Eaux at Paris.

On this side of the Atlantic also I am indebted to Hiram F. Mills, C.E., under whose direction I had the privilege of conducting for nearly five years the Lawrence experiments on filtration; to Profs. Sedgwick and Drown for the numerous suggestions and friendly criticisms, and to the latter for kindly reading the proof of this volume; to Mr. G. W. Fuller for full information in regard to the more recent Lawrence results; to Mr. H. W. Clark for the laborious examination of the large number of samples of sands used in actual filters and mentioned in this volume; and to Mr. Desmond FitzGerald for unpublished information in regard to the results of his valuable experiments on filtration at the Chestnut Hill Reservoir, Boston.

ALLEN HAZEN.

BOSTON, April, 1895.

FILTRATION OF PUBLIC WATER-SUPPLIES.

CHAPTER I.

INTRODUCTION.

The rapid and enormous development and extension of water-works in every civilized country during the past forty years is a matter which deserves our most careful consideration, as there is hardly a subject which more directly affects the health and happiness of almost every single inhabitant of all cities and large towns.

Considering the modern methods of communication, and the free exchange of ideas between nations, it is really marvellous how each country has met its problems of water-supply from its own resources, and often without much regard to the methods which had been found most useful elsewhere. England has secured a whole series of magnificent supplies by impounding the waters of small streams in reservoirs holding enough water to last through dry periods, while on Continental Europe such supplies are hardly known. Germany has spent millions upon millions in purifying turbid and polluted river-waters, while France and Austria have striven for mountain-spring waters and have built hundreds of miles of costly aqueducts to secure them. In the United States an abundant supply of some liquid has too often been the objective point, and the efforts have been most successful, the American works being entirely unrivalled in the volumes of their supplies. I do not wish to imply that quality has been entirely neglected in our country, for many cities and towns have seriously and successfully studied their problems, with the result that there are hundreds of water-supplies in the United States which will compare favorably upon any basis with supplies in any part of the world; but on the other hand it is equally true that there are hundreds of other cities, including some among the largest in the country, which supply their citizens with turbid and unhealthy waters which cannot be regarded as anything else than a national disgrace and a menace to our prosperity.

One can travel through England, Belgium, Holland, Germany, and large portions of other European countries and drink the water at every city visited without anxiety as to its effect upon his health. It has not always been so. Formerly European capitals drank water no better than that so often dispensed now in America. It is but two years since Germany's great commercial centre, Hamburg, having a water-supply essentially like those of Philadelphia, Pittsburg, Cincinnati, St. Louis, New Orleans, and a hundred other American cities, paid a penalty in one month of eight thousand lives for its carelessness. The lesson was a dear one, but it was not wasted. Hamburg now has a new and wholesome supply, and other German cities the qualities of whose waters were open to question have been forced to take active measures to better their conditions. We also can learn something from their experience.

There are three principal methods of securing a good water-supply for a large city. The first consists of damming a stream from an uninhabited or but sparsely inhabited watershed, thus forming an impounding reservoir. This method is extensively used in England and in the United States. In the latter most of the really good and large supplies are so obtained. It is only applicable to places having suitable watersheds within a reasonable distance, and there are large regions where, owing to geological and other conditions, it cannot be applied. It is most useful in hilly and poor farming countries, as in parts of England and Wales, in the Atlantic States, and in California. It cannot be used to any considerable extent in level and fertile countries which are sure to be or to become densely populated, as is the case with large parts of France and Germany and in the Middle States.

The second method is to secure ground-water, that is, spring or well water, which by its passage through the ground has become thoroughly purified from any impurities which it may have contained. This was the earliest and is the most widely used method of securing good water. It is specially adapted to small supplies. Under favorable geological conditions very large supplies have been obtained in this manner. In Europe Paris, Vienna, Budapest, Munich, Cologne, Leipzig, Dresden, a part of London, and very many smaller places are so supplied. This method is also extensively used in the United States for small and medium-sized places, and deserves to be most carefully studied, and used whenever possible, but is unfortunately limited by geological conditions and cannot be used except in a fraction of the cases where supplies are required. No ground-water supplies yet developed in the United States are comparable in size to those used in Europe.

The third process of securing a good water-supply is by means of filtration of surface waters which would otherwise be unsuitable for domestic purposes. The methods of filtration, which it is the purpose of this volume to explain, are beyond the experimental stage; they are now applied to the purification of the water-supplies of European cities with an aggregate population of at least 20,000,000 people. In the United States there are possibly one two hundredth as many people so supplied, and in most of these cases the filters are of quite recent introduction. As far as the bulk of our people, even those in official positions, are concerned, it is no exaggeration to say that the modern methods of filtration are unknown on this side of the Atlantic. In Europe filtration has been practised with continually improving methods since 1839, and the process has steadily received wider and wider application. It has been most searchingly investigated in its hygienic relations, and has been repeatedly found to be a most valuable aid in reducing mortality. The conditions under which satisfactory results can be obtained are now tolerably well known, so that filters can be built in the United States with the utmost confidence that the result will not be disappointing.

The cost of filtration, although considerable, is not so great as to put it beyond the reach of American cities. It may be roughly estimated that the cost of filtration, with all necessary interest and sinking funds, will add 10 per cent to the average cost of water as at present supplied.

It may be confidently expected that when the facts are better understood and realized by the American public, we shall abandon the present filthy and unhealthy habit of drinking polluted river and lake waters, and shall put the quality as well as the quantity of our supplies upon a level not exceeded by those of any country.

CHAPTER II.

CONTINUOUS FILTERS AND THEIR CONSTRUCTION.

FILTRATION of water consists in passing it through some substance which retains or removes some of its impurities. In its simplest form filtration is a straining process, and the results obtained depend upon the fineness of the strainer, and this in turn is regulated by the character of the water and the uses to which it is to be put. Thus in the manufacture of paper an enormous volume of water is required free from particles which, if they should become imbedded in the paper, would injure its appearance or texture. Obviously for this purpose the removal of the smaller particles separately invisible to the unaided eye, and thus not affecting the appearance of the paper, and the removal of which would require the use of a finer filter at increased expense, would be a simple waste of money. When, however, a water is to be used for a domestic water supply and transparency is an object, the still finer particles which would not show themselves in paper, but which are still able, in bulk, to render a water turbid, should be as far as possible removed, thus necessitating a finer filter ; and, when there is reason to think that the water contains the germs of disease, the filter must be fine enough to remove with certainty those organisms so extraordinarily small that millions of them may exist in a glass of water without imparting a visible turbidity.

It is now something over half a century since the first successful attempts were made to filter public water-supplies, and there are now hundreds of cities supplied with clear, healthy, filtered water. (Appendix IV.) While the details of the filters

FILTRATION OF PUBLIC WATER-SUPPLIES.

used in different places present considerable variations, the general form is, in Europe at least, everywhere the same. The most important parts of a filter are shown by the accompanying sketch,

FIG. 1.—SKETCH SHOWING GENERAL ARRANGEMENT OF FILTER PLANTS.

in which the dimensions are much **exaggerated**. The raw water is taken from the **river** into a settling-basin, where the heaviest mud **is allowed to settle.** In the case of lake and pond waters the settling-tank **is dispensed** with, but it is essential for turbid river-water, **as otherwise the** mud clogs the filter too rapidly. The partially **clarified water then passes** to the filter, which consists **of a horizontal layer of rather fine** sand supported by gravel **and underdrained, the whole** being enclosed in a suitable basin or tank. **The water in** passing through the sand leaves behind upon the sand grains the extremely small particles which were too fine to **settle out in** the settling-basin, and is quite clear as it goes from the gravel to the drains and the pumps, which forward it to the reservoir **or city.**

The passages between the **grains of** sand through which the water must pass **are** extremely **small.** If the sand grains were spherical and $\frac{1}{20}$ of an inch in diameter, the openings would only allow **the passage of other** spheres $\frac{1}{320}$ of an inch in diameter, **and with** actual irregular **sands much finer** particles are held **back. As a** result the **coarser matters in the** water are retained on the **surface of the sand,** where **they** quickly form a layer of sediment, **which itself** becomes a filter much finer than **the sand alone, and which is capable of** holding back under suitable **conditions even the bacteria of the passing** water. The **water which passes before this takes place** may be less perfectly filtered, but even then, the filter may be so operated that nearly all of the bacteria will be deposited in the sand and not allowed to pass through into the effluent.

As the sediment layer increases in thickness with continued filtration, increased pressure is required to drive the desired volume of water through its pores, which are ever becoming smaller and reduced in number. When the required quantity of water will no longer pass with the maximum pressure allowed, it is necessary to remove, by scraping, the sediment layer, which should not be more than an inch deep. This layer contains most of the sediment, and the remaining sand will then act almost as new sand would do. The sand removed may be washed for use again, and eventually replaced when the sand layer becomes too thin by repeated scrapings. These operations require that the filter shall be temporarily out of use, and as water must in general be supplied without intermission, a number of filters are built together, so that any of them can be shut out without interfering with the action of the others.

The arrangement of filters in relation to the pumps varies with local conditions. With gravity supplies the filters are usually located below the storage reservoir, and, properly placed, involve only a few feet loss of head.

In the case of tidal rivers, as at Antwerp and Rotterdam, the quality of the raw water varies with the tide, and there is a great advantage in having the settling-basins low enough so that a whole day's supply can be rapidly let in when the water is at its best, without pumping. At Antwerp the filters are higher, and the water is pumped from the settling basins to them, and again from the reservoir receiving the effluents from the filters to the city. In several of the London works (East London, Grand Junction, Southwark and Vauxhall, etc.) the settling-basins are lower than the river, and the filters are still lower, so that a single pumping suffices, that coming between the filter and the city, or elevated distributing reservoir.

In many other English filters and in most German works the settling-basins and filters are placed together a little higher than the river, thus avoiding at once trouble from floods and cost for excavation. The water requires to be pumped twice, once before and once after filtration. At Altona the settling-basins and filters are placed upon a hill, to which the raw Elbe water is pumped, and from which it is supplied to the city after filtration by gravity without further pumping. The location of the works in this case is said to have been determined by the location of a bed of sand suitable for filtration on the spot where the filters were built.

When two pumpings are required they are frequently done, especially in the smaller places, in the same pumping-station, with but one set of boilers and engines, the two pumps being connected to the same engine. The cost is said to be only slightly greater than that of a single lift of the same total height. In very large works, as at Berlin and Hamburg and some of the London companies, two separate sets of pumping machinery involve less extra cost relatively than would be the case with smaller works.

SEDIMENTATION-BASINS.

Kirkwood * found in 1866 that sedimentation-basins were essential to the successful treatment of turbid river-waters, and subsequent experience has not in any way shaken his conclusion. The German works visited by him, Berlin (Stralau) and Altona, were both built by English engineers, and their settling-basins did not differ materially from those of corresponding works in England. Since that time, however, there has been a well-marked tendency on the part of the German engineers to use smaller, while the English engineers have used much larger sedimentation-basins, so that the practices of the two countries are

---

* Filtration of River Waters. Van Nostrand & Co., 1869.

now widely separated, the difference no doubt being in part at least due to local causes.

Kirkwood found sedimentation-basins at Altona with a capacity of 2¼ times the daily supply. In 1894 the same basins were in use, although the filtering area had been increased from 0.82 acre to 2.20 acres, and still more filters were in course of construction, and the average daily quantity of water had increased from 600,000 to 4,150,000 gallons in 1891-2, or more than three times the capacity of the sedimentation-basins. In 1890 the depth of mud deposited in these basins was reported to be two feet deep in three months. At Stralau in Berlin, also, in the same time the filtering area was nearly doubled without increasing the size of the sedimentation-basins, but the Spree at this point has such a slow current that it forms itself a natural sedimentation-basin. At Magdeburg on the Elbe works were built in 1876 with a filtering area of 1.92 acres, and a sedimentation-basin capacity of 11,300,000 gallons, but in 1894 half of the latter had been built over into filters, which with two other filters gave a total filtering surface of 3.90 acres, with a sedimentation-basin capacity of only 5,650,000 gallons. The daily quantity of water pumped for 1891-2 was 5,000,000 gallons, so that the present sedimentation-basin capacity is about equal to one day's supply, or relatively less than a third of the original provision. The idea followed is that most of the particles which will settle at all will do so within twenty-four hours, and that a greater storage capacity may allow the growth of algæ, and that the water may deteriorate rather than improve in larger tanks.

At London, on the other hand, the authorities consider a large storage capacity for unfiltered water as one of the most important conditions of successful filtration, the object however, being perhaps as much to secure storage as to allow sedimentation. In 1893 thirty-nine places were reported upon the Thames and the Lea which were giving their sewage systematic treatment before discharging it into the streams from which London's water is drawn. These sewage treatments are, with hardly an exception, dry-weather treatments, and as soon as there is a considerable storm crude sewage is discharged into the rivers at every point. The rivers are both short, and are quickly flooded, and afterwards are soon back in their usual condition. At these times of flood, the raw water is both very turbid and more polluted by sewage than at other times, and it is the aim of the authorities to have the water companies provide reservoir capacity enough to carry them through times of flood without drawing any water whatever from the rivers. This obviously involves much more extensive reservoirs than those used in Germany, and the companies actually have large basins and are still adding to them. The storage capacities of the various companies vary from 3 to 18 times the respective average daily supplies, and together equal 9 times the total supply.

In case the raw water is taken from a lake or a river at a point where there is but little current, as in a natural or artificial pond, sedimentation-basins are unnecessary. This is the case at Zürich (lake water), at Berlin when the rivers Havel and Spree spread into lakes, at Tegel and Müggel, and at numerous other works.

SIZE OF FILTER-BEDS.

The total area of filters required in any case is calculated from the quantity of water required, the rate of filtration, and an allowance for filters out of use while being cleaned. To prevent interruptions of the supply at times of cleaning, the filtering area is divided into beds which are operated separately, the number and size of the beds depending upon local conditions. The cost per acre is decreased with large beds on account of there being less wall or embankment required, while, on the other hand, the convenience of operation may suffer, especially in small works. It is also frequently urged that with large filters it is difficult or impossible to get an even rate of filtration over the entire area owing to the frictional resistance of the underdrains for the more distant parts of the filter. A discussion of this point is given in Chapter III, page 37. At Hamburg, where the size of the single beds, 1.88 acres each, is larger than at any other place, it is shown that there is no serious cause for anxiety; and even if there were, the objectionable resistance could be still farther reduced by a few changes in the under-drains. The sizes of filter-beds used at a large number of places are given in Appendix IV.

At a number of places having severe winters, filters are vaulted over as a protection from cold, and in the most important of these, Berlin, Warsaw, and St. Petersburg, the areas of the single beds are nearly the same, namely, from 0.52 to 0.59 acre. The works with open filters at London (seven companies), Amsterdam, and Breslau have filter-beds from 0.82 to 1.50 acres each. Liverpool and Hamburg alone use filters with somewhat larger areas. Large numbers of works with both covered and open filters have much smaller beds than these sizes, but generally this is to avoid too small a number of divisions in a small total area, although such works have sometimes been extended with the growth of the cities until they now have a considerable number of very small basins.

FORM OF FILTER-BEDS.

The form and construction of the filter-beds depend upon local conditions, the foundations, and building materials available, the principles governing these points being in general the same as for the construction of ordinary reservoirs. The bottoms require to be made water-tight, either by a thin layer of concrete or by a pavement upon a puddle layer. For the sides either masonry walls or embankments are used, the former saving space, but being in general more expensive in construction. Embankments must, of course, be substantially paved near the water-line to withstand the action of ice, and must not be injured by rapid fluctuations in the water-levels in the filters.

Failure to make the bottoms water-tight has perhaps caused more annoyance than any other single point. With a leaky bottom there is either a loss of water when the water in the filters is higher than the ground-water, or under reverse conditions, the ground-water comes in and mixes with the filtered water, and the latter is rarely improved and may be seriously damaged by the admixture. And with very bad conditions water may pass from one filter to another, with the differences in pressure always existing in neighboring filters, with most unsatisfactory results.

COVERS FOR FILTERS.

The filters in England and Holland are built open, without protection from the weather. In Germany the filters first built were also open, but in the colder climates more or less difficulty was experienced in keeping the filters in operation in cold weather. An addition to the Berlin filters, built in 1874, was covered with masonry vaulting, over which several feet of earth were placed, affording a complete protection against frost. The filters at Magdeburg built two years later were covered in the same way, and since that time covered filters have been built at perhaps a dozen different places.

It was found at Berlin that, owing to the difficulty of properly cleaning the open filters in winter, it was impossible to keep the usual proportion of the area in effective service, and as a result portions of the filters were greatly overtaxed during prolonged periods of cold weather. This resulted in greatly decreased bacterial efficiency, the bacteria in March, 1889, reaching 3000 to 4000 per cc. (with 100,000 in the raw water), although ordinarily the effluent contained less than 100. An epidemic of typhoid fever followed, and was confined to that part of the city supplied from the Stralau works, the wards supplied from the covered Tegel filters remaining free from fever. **Open** filters have since been abandoned **in Berlin.**

At Altona also, where the water **is taken** from an excessively polluted source, decreased bacterial efficiency has repeatedly resulted in winter, and the occasional epidemics of typhoid fever in that city, which have invariably **come in** winter, appear to have been directly **due** to the effect of cold **upon the** open filters. The city has just extended the open filters, **and** hopes with an increased reserve area to avoid the difficulty in future without resource to **covered** filters. (See Appendices II and **VII.**)

Brunswick, Lübeck, and Frankfort on Oder with cold winters have **open filters, but draw their** water-supplies from **less** polluted sources, and **have** thus far escaped the **fate of** Berlin and Altona. **The new** filters at Hamburg also are open. At Zürich, where open and covered filters were long used side by side, the covered filters were much more satisfactory, and the old open filters have recently been vaulted over.

Königsberg originally built **open filters, but was** afterward obliged to cover them, **on** account of the severe winters; and at Breslau, where open filters have long been used, the recent additions are vaulted over.

The fact that inferior efficiency **of** filtration results with open filters during prolonged and severe winter weather is generally admitted, although there is some doubt as to the **exact** way in which the disturbance is caused. In some works I am informed that in cutting the ice around the edges of the filter and repeatedly piling the loose pieces upon the floating cake, the latter eventually becomes so thickened **at** the sides that the projecting lower corners actually touch the sand, with the fluctuating levels which often prevail in these works, and that in this way the sediment layer upon the top of the sand is broken and the water rapidly passes without adequate purification at the points of dis**turbance.**

This theory is, however, inadequate to account for many cases where such an accumulation of ice is not allowed. In these cases the poor work is not obtained until after the filters have been scraped. The sand apparently freezes slightly while the water is off, and when water is brought back and filtration resumed, normal results are for some reason not again obtained for a time.

In addition to the poorer work from open filters in cold weather, the cost of removing the ice adds materially to the operating expenses, and in very cold climates would in itself make covers advisable.

I have arranged the European filter plants, in regard to which I have sufficient information, in the table on page 15, in the order of the normal mean January temperatures of the respective places. This may not be an ideal criterion of the necessity of covering filters, but it is at least approximate, and in the absence of more detailed comparisons it will serve to give a good general idea of the case. I have not found a single case where covered filters are used where the January temperature is 32° F. or above. In some of these places some trouble is experienced in unusually cold weather, but I have not heard of any very serious difficulty or of any talk of covering filters at these places except at Rotterdam, where a project for covering was being discussed.

Those places having January temperatures below 30° experience a great deal of difficulty with open filters; so much so, that covered filters may be regarded as necessary for them, although it is possible to keep open filters running with decreased efficiency and increased expense by freely removing the ice, with January temperatures some degrees lower.

Where the mean January temperature is 30° to 32° F. there is room for doubt as to the necessity of covering filters, but, judging from the experience of Berlin and Altona, the covered filters are much safer at this temperature.

CONTINUOUS FILTERS AND THEIR CONSTRUCTION. 15

TABLE OF PLACES HAVING OPEN AND COVERED FILTERS.
ARRANGED ACCORDING TO THE MEAN JANUARY TEMPERATURES.

Normal Mean January Temperature. Degrees F	Place.	Kind of Filters and Results.
37–40°	All English cities	Open filters only are used, and no great difficulty with ice is experienced.
33–35°	Cities in Holland	All filters are open, and there is little serious trouble with ice; but at Amsterdam and Rotterdam the bacteria in effluents are said to be higher in winter than at other times.
32°	Bremen	Open filters.
31°	Altona	Much difficulty with ice in open filters (see Appendices II and VII).
31°	Brunswick	Open filters.
31°	Hamburg	" "
31°	Lubeck	" "
31°	Berlin	Open filters were formerly used, but owing to decreased efficiency in cold weather they have been abandoned for covered ones.
31°	Magdeburg	Covered filters, but a recent addition is not covered.
30°	Frankfort on Oder	Open filters.
30°	Stuttgart	Part of the filters are covered.
30°	Stettin	" " " "
29°	Zurich	Covered filters were much the most satisfactory, and the open ones were covered in 1894. The raw water has a temperature of 35°.
29°	Liegnitz	Open filters.
29°	Breslau	Open filters have been used, but recent additions are covered.
29°	Budapest	Covered filters only.
29°	Posen	" " "
26°	Königsberg	The original filters were open, but it was found necessary to cover them.
24°	Warsaw	Covered filters only.
16°	St. Petersburg	" " "

In case the raw water was drawn from a lake at a depth where its minimum temperature was above 32°, which is the temperature which must ordinarily be expected in surface-waters in winter, open filters might be successfully used in slightly colder places.

The covers are usually of brick or concrete vaulting supported by pillars at distances of 11 to 15 feet in each direction, the whole being covered by 2 or 3 feet of earth; and the top can be laid out as a garden if desired. Small holes for the admission of air and light are usually left at intervals. The thickness of the masonry and the sizes of the pillars used in some of the earlier German vaultings are unnecessarily great, and some of the newer works are much lighter. For American use, vaulting like that used for the Newton, Mass., covered reservoir* should be amply strong.

Roofs have been used at Königsberg, Posen, and Budapest instead of the masonry vaulting. They are cheaper, but do not afford as good protection against frost, and even with great care some ice will form under them.

Provision must be made for entering the filters freely to introduce and remove sand. This is usually accomplished by raising one section of vaulting and building a permanent incline under it from the sand line to a door above the high-water line in the filter.

The cost of building covered filters is said to average fully one half more than open filters.

Among the incidental advantages of covered filters is that with the comparative darkness there is no tendency to algæ growths on the filters in summer, and the frequency of scraping is therefore somewhat reduced. At Zurich, in 1892, where both covered and open filters were in use side by side, the periods between scrapings averaged a third longer in the covered than in the open filters.

It has been supposed that covered filters kept the water cool in summer and warm in winter, but owing to the large volume of water passing, the change in temperature in any case is very slight; Frühling found that even in extreme cases a change of over 3° F. in either direction is rarely observed.

---

* Annual Report of Albert F. Noyes, City Engineer for 1891.

At Berlin, where open and covered filters were used side by side at Stralau for twenty years, it was found that, bacterially, the open filters were, except in severe winter weather, more efficient. It was long supposed that this was caused by the sterilizing action of the sunlight upon the water in the open filters. This result, however, was not confirmed elsewhere, and it was finally discovered, in 1893, that the higher numbers were due to the existence of passages in corners on the columns of the vaulted roof and around the ventilators for the underdrains, through which, practically, unfiltered water found its way into the effluent. This at once removes the evidence in favor of the superior bacterial efficiency of open filters and suggests the necessity of preventing such passages. The construction of a ledge all around the walls and pillars four inches wide and a little above the gravel, as shown in the sketch, might be useful in this way, and the slight lateral movement of the water in the sand above would be of no consequence. The sand would evidently make a closer joint with the horizontal ledge than with the vertical wall.

In regard to the probable requirement or advisability of covers for filters in the United States, I judge, from the European experience, that places having January temperatures below the freezing-point will have considerable trouble from open filters, and would best have covered filters. Places having higher winter temperatures will be able to get along with the ice which may form on open filters, and the construction of covers would hardly be advisable except under exceptional local conditions, as, for instance, with a water with an unusual tendency to algæ growths.

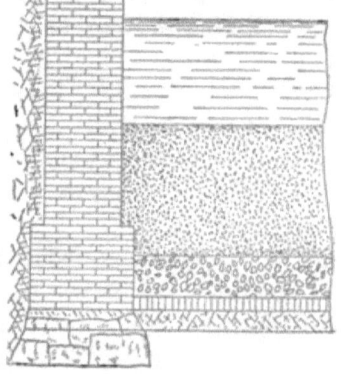

FIG. 2.

I have drawn a line across a map of the United States on this basis (shown by the accompanying plate) and it would appear that places far north of the line would require covered filters, and that those south of it would not, while for the places in the immediate vicinity of the line (comparable to Hamburg and Altona) there is room for discussion.

In case open filters are built near or north of this line, I would suggest that plenty of space between and around the filters for piling up ice in case of necessity may be found advantageous, and that a greater reserve of filtering area for use in emergencies should be provided than would be considered necessary with vaulted filters or with open filters in a warmer climate.

CHAPTER III.

FILTERING MATERIALS.

SAND.

THE sand used for filtration may be obtained from the seashore, from river-beds or from sand-banks. It consists mainly of sharp quartz grains, but may also contain hard silicates. As it occurs in nature it is frequently mixed with clayey or other fine particles, which must be removed from it by washing before it is used. Some of the New England sands, however, as that used for the Lawrence City filter, are so clean that washing would be superfluous.

The grain size of the sand best adapted to filtration has been variously stated at from $\frac{1}{3}$ to 1 mm., or from 0.013 to 0.040 inch. The variations in the figures, however, are due more to the way that the same sand appears to different observers than to actual variations in the size of sands used, which are but a small fraction of those indicated by these figures.

As a result of experiments made at the Lawrence Experiment Station * we have a standard by which we can definitely compare various sands. The size of a sand-grain is uniformly taken as the diameter of a sphere of equal volume, regardless of its shape. As a result of numerous measurements of grains of Lawrence sands, it is found that when the diameter, as given above, is 1, the three axes of the grain, selecting the longest possible and taking the other two at right angles to it, are, on an average, 1.38, 1.05, and 0.69, respectively and the mean diameter is equal to the cube root of their product.

---

* Rept. Mass. State Board of Health, 1892, p. 541. See Appendix III.

It was also found that in mixed materials containing particles of various sizes the water is forced to go around the larger particles and through the finer portions which occupy the intervening spaces, so that it is the finest portion which mainly determines the character of the sand for filtration. As a provisional basis which best accounts for the known facts, the size of grain such that 10 per cent by weight of the particles are smaller and 90 per cent larger than itself, is considered to be the *effective size*. The size so calculated is uniformly referred to in speaking of the size of grain in this work.

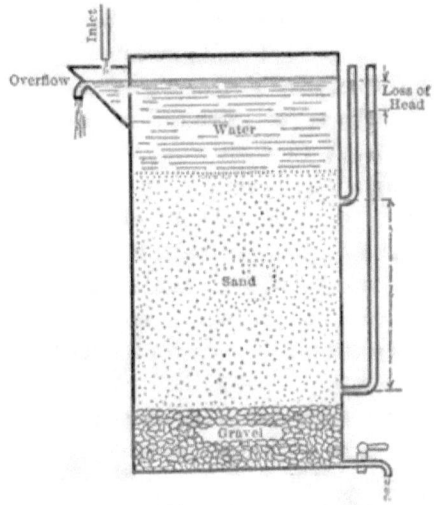

FIG. 3.—APPARATUS USED FOR MEASURING THE FRICTION OF WATER IN SANDS.

Another important point in regard to a material is its degree of uniformity—whether the particles are mainly of the same size or whether there is a great range in their diameters. This is shown by the *uniformity coefficient*, a term used to designate the ratio of the size of grain which has 60 per cent of the sample finer than itself to the size which has 10 per cent finer than itself.

FILTERING MATERIALS.

The frictional resistance of sand to water when closely packed, with the pores completely filled with water and in the entire absence of clogging, was found to be expressed by the formula

$$v = cd^2 \frac{h}{l}\left(\frac{t\ \text{Fah.} + 10°}{60}\right),$$

where v is the velocity of the water in meters daily in a solid column of the same area as that of the sand, or approximately in million gallons per acre daily;

c is a constant factor which present experiments indicate to be approximately 1000;

d is the effective size of sand grain in millimeters;

h is the loss of head (Fig. 3);

l is the thickness of sand through which the water passes;

t is the temperature (Fahr.).

The formula can only be used for sands with uniformity coefficients below 5 and effective sizes from 0.10 to 3.00 mm., and with the coarser materials only for moderately low rates. The following table shows the rates, in million gallons per acre daily, at which water will pass through sands of different sizes.

TABLE SHOWING RATE AT WHICH WATER WILL PASS THROUGH DIFFERENT-SIZED SANDS WITH VARIOUS HEADS AT A TEMPERATURE OF 50°.

$\frac{h}{l}$	Effective Size in Millimeters 10 per cent finer than:							
	0.10	0.20	0.30	0.35	0.40	0.50	1.00	3.00
	Million	Gallons	per Acre	daily.				
.001	.01	.04	.10	.13	.17	.27	1.07	9.63
.005	.05	.21	.48	.65	.85	1.34	5.35	48.15
.010	.11	.43	.96	1.31	1.71	2.67	10.70	96.30
.050	.54	2.14	4.82	6.55	8.55	13.40	53.50	
.100	1.07	4.28	9.63	13.10	17.10	26.70	107.00	..
1.000	10.70	42.80	96.30	131.00	171.00	267.00	....	

FILTRATION OF PUBLIC WATER-SUPPLIES.

RELATIVE QUANTITIES OF WATER PASSING **AT** DIFFERENT TEMPERATURES.

32°....0.70	44°....0.90	56°....1.10	68°....1.30
35°....0.75	47°....0.95	59°....1.15	71°....1.35
38°....0.80	50°....1.00	62°....1.20	74°....1.40
41°....0.85	53°....1.05	65°....1.25	77°....1.45

SANDS USED IN EUROPEAN FILTERS.

To secure definite information in regard to the qualities of the sands actually used in filtration, a large number of European works were visited in 1894, and samples of sand were collected for analysis. These samples were examined at the Lawrence Experiment Station by Mr. H. W. Clark, the author's method of analysis described in Appendix III being used. In the following table, for the sake of compactness, only the leading points of the analyses, namely, effective size, uniformity coefficient, and albuminoid ammonia, are given. On page 26 full analyses of some samples from a few of the leading works are given.

The English and most of the German sands are washed, even when entirely new, before being used, to remove fine particles. At Breslau, however, sand dredged from the river Oder is used in its natural state, and new sand is used for replacing that removed by scraping. At Budapest, Danube sand is used in the same way, but with a very crude washing, and it is said that only new unwashed sand is used at Warsaw.

In Holland, so far as I learned, no sand is washed, but new sand is always used for refilling. At most of the works visited dune-sand with an effective size of only 0.17 to 0.19 mm. is used, and this is the finest sand which I have ever found used for water filtration on a large scale. It should be said, however, that the waters filtered through these fine sands are fairly clear before filtration, and are not comparable to the turbid river-waters often

ANALYSES OF SANDS USED IN WATER FILTRATION.

Source.	Effective Size: 10% Finer than (Millimeters).	Uniformity Coefficient.	Albuminoid Ammonia. Parts in 100,000.	Remarks.
London, E. London Co.	0.44	1.8	0.45	New sand, never used or washed.
" "	0.39	2.1	26.20	Dirty sand, very old.
" "	0.37	2.0	8.60	Same, washed by hand.
" Grand Junc...	0.26	1.9	1.90	Sand from rough filter.
" " "	0.40	3.5	10.00	Old sand in final filter.
" " "	0.41	3.7	2.70	Freshly washed old sand.
" Southw'k & V	0.38	3.5	5.00	" " " "
" "	0.30	1.8	2.80	Freshly washed new sand.
" Lambeth......	0.36	2.3	2.60	Freshly washed old sand.
" "	0.36	2.4	0.35	New unused sand, washed.
" "	0.25	1.7	0.70	New extremely fine sand.
" Chelsea.....	0.36	2.4	2.10	Freshly washed old sand.
Middlesborough.......	0.42	1.6	17.60	Dirty sand, ordinary scraping.
"	0.43	1.6	7.30	Same, after washing.
Birmingham........	0.29	1.9	33.20	Dirty sand.
"	0.29	1.9	7.20	Sand below surface of filter.
Reading..............	0.30	2.5	4.00	Dirty sand.
"	0.22	2.0	1.50	Same, after washing.
Antwerp.............	0.38	1.6	7.80	Dirty sand.
"	0.39	1.6	3.40	Same, after washing.
Hamburg.............	0.28	2.5	8.50	Dirty sand.
"	0.31	2.3	0.80	Same, after washing.
"	0.34	2.2	7.90	Dirty sand, another sample.
"	0.30	2.0	0.90	Same, after washing drums.
"	0.34	2.3	1.50	" " " ejectors.
Altona...............	0.32	2.0	9.00	Dirty sand, old filters.
"	0.37	2.0	1.50	Same, after washing.
"	0.33	2.8	0.50	Washed sand for new filters.
Berlin, Stralau...	0.33	1.9	12.20	Dirty sand-pile.
" "	0.35	1.7	4.50	Filter No. 6, 3″ below surface.
" "	0.34	1.7	6.30	" " 7 " "
" "	0.35	1.7	4.00	" " 10 " "
" Tegel..........	0.38	1.6	11.00	Dirty sand, old filters.
" "	0.38	1.5	2.80	Same, after washing, old filters.
" "	0.35	1.6	3.20	" " " new "
" Müggel........	0.35	1.8	0.80	Sand from filters below surface.
" "	0.33	2.0	6.30	Dirty sand, ordinary scraping.
" "	0.34	2.0	15.30	" " another sample.
Charlottenburg......	0.40	2.3	7.20	" "
Chemnitz............	0.35	2.6	0.20	New sand not yet used.
Magdeburg...........	0.39	2.0	9.50	Dirty sand.
"	0.40	2.0	2.80	Same, after washing.
Breslau..............	0.39	1.8	1.40	Normal new sand.
Budapest.............	0.20	2.0	0.80	New washed Danube sand.
Zurich...............	0.28	3.2	6.20	Dirty sand.
"	0.30	3.1	1.50	Same, after washing.
Hague................	0.19	1.6	0.70	Dune-sand used for filtration.

ANALYSES OF SANDS USED IN WATER FILTRATION.—*Continued.*

Source.	Effective Size; 10% Finer than (Millimeters).	Uniformity Coefficient.	Albuminoid Ammonia. Parts in 100,000.	Remarks.
Schiedam............	0.18	1.6	5.60	Dune-sand used for filtration dirty.
"	0.31	1.5	13.50	River-sand; dirty.
Amsterdam..........	0.17	1.6	2.40	Dune-sand.
Rotterdam...........	0.34	1.5	2.30	River-sand; new.
Liverpool, Rivington..	0.43	2.0	0.76	Sand from bottom of filter.
" "	0.32	2.5	1.00	New sand unwashed and unscreened.
" "	0.43	2.7	4.10	Washed sand which has been in use 30 to 40 years.
" Owesty......	0.30	2.6	9.40	Dirty sand.
" "	0.31	4.7	2.20	Same, after washing.

NOTE.—It is obvious that in case the sands used at any place are not always of the same character, as is shown to be the case by different samples from some of the works, the examination of such a limited number of samples as the above from each place is entirely inadequate to establish accurately the sizes of sand used at that particular place, or to allow close comparisons between the different works, and for this reason no such comparisons will be made. The object of these investigations was to determine the sizes of the sands commonly used in Europe, and, considering the number and character of the different works represented, it is believed that the results are ample for this purpose.

filtered elsewhere, and their tendency to choke the filters is consequently much less. At Rotterdam and Schiedam, where the raw water is drawn from the Maas, as the principal stream of the Rhine is called in Holland, river-sand of much larger grain size is employed. It is obtained by dredging in the river and is never washed, new sand always being employed for refilling.

The average results of the complete analyses of sands from ten leading works are shown in the table on page 26. These figures are the average of all the analyses for the respective places, except that one sample from the Lambeth Co., which was not a representative one, was omitted.

The London companies were selected for this comparison both on account of their long and favorable records in filtering the polluted waters of the Thames and Lea, and because they are subject to close inspection; and there is ample evidence that the filtration obtained is good—evidence which is often lacking in the smaller and less closely watched works. For the German works Altona was selected because of its escape from cholera in 1892, due to the efficient action of its filters, and Stralau because of its long and favorable record when filtering the much-polluted Spree water. These two works also have perhaps contributed more to the modern theories of filtration than all the other works in existence. The remaining works are included because they are comparatively new, and have been constructed with the greatest care and attention to details throughout, and the results obtained are most carefully recorded.

The averages show the effective size of the English sands to be slightly greater than that of the German sands—0.37 instead of 0.34 mm.—but the difference is very small. The entire range for the ten works is only from 0.31 to 0.40 mm., and these may be taken as the ordinary limits of effective size of the sands employed in the best European works. The average for the other sixteen works given above, including dune-sands, is 0.31 mm., or, omitting the dune-sands, 0.34 mm.

Turning to the circumstances which influence the selection of the sand size, we find that both the quality of the effluent obtained by filtration and the cost of filtration depend upon the size of the sand-grains.

With a fine sand the sediment layer forms more quickly and the removal of bacteria is more complete, but, on the other hand, the filter clogs quicker and the dirty sand is more difficult to wash, so that the expense is increased.

EFFECT OF SIZE OF GRAIN UPON EFFICIENCY OF FILTRATION.

It is frequently stated that it is only the sediment layer which performs the work of filtration, and that the sand which supports it plays hardly a larger part than does the gravel which

FILTRATION OF PUBLIC WATER-SUPPLIES.

TABLE SHOWING THE AVERAGE PER CENT OF THE GRAINS FINER THAN VARIOUS SIZES IN SANDS FROM LEADING WORKS.

	Per Cent by Weight Finer than							
	0.106 mm.	0.186 mm.	0.316 mm.	0.46 mm.	0.93 mm.	2.04 mm.	3.89 mm.	5.89 mm.
East London	0.2	0.5	3.6	22.2	69.7	89.8	95.0	99.0
Grand Junction	0	0.2	3.1	17.4	47.1	68.2	84.7	93.6
Southwark and Vauxhall	...	0.7	8.0	34.1	69.7	83.5	90.0	94.0
Lambeth	0	0.5	5.5	26.6	63.0	79.2	88.0	94.3
Chelsea	0	0.1	5.0	28.6	63.0	76.7	86.0	93.6
Hamburg	0.2	1.5	10.9	33.2	74.4	95.7	99.5	
Altona	0.1	1.1	7.8	28.7	72.1	92.1	95.8	
Stralau	...	0.3	7.0	37.3	86.9	95.4	97.6	
Tegel	...	0.2	4.5	35.4	94.3	98.5	99.1	
Müggel	0.1	0.5	7.9	33.6	79.7	94.3	98.5	
Average of all	0.06	0.56	6.33	29.71	71.99	87.34	93.42	(97.45)

AVERAGE EFFECTIVE SIZE, UNIFORMITY COEFFICIENT, AND ALBUMINOID AMMONIA IN SANDS FROM TEN LEADING WORKS.

I. LONDON FILTERS.

	Effective Size; 10% Finer than (Millimeters).	Uniformity Coefficient.	Albuminoid Ammonia.	
			Dirty Sand.	Washed Sand.
East London	0.40	2.0	26.00	8.60
Grand Junction	0.40	3.6	10.00	2.70
Southwark and Vauxhall	0.34	2.5	.....	3.90
Lambeth	0.36	2.4	.....	2.60
Chelsea	0.36	2.4	.....	2.10
Average	0.37	2.6	18.00	3.98

II. GERMAN WORKS.

	Effective Size; 10% Finer than (Millimeters).	Uniformity Coefficient.	Albuminoid Ammonia.	
			Dirty Sand.	Washed Sand.
Stralau	0.34	1.7	12.20	4.00
Tegel	0.37	1.6	11.00	3.00
Müggel	0.34	2.0	10.80	0.80
Altona	0.34	2.3	9.00	1.50
Hamburg	0.31	2.3	8.20	1.07
Average	0.34	2.0	10.25	2.07

carries the sand, and under some circumstances this is undoubtedly the case. Nevertheless sand in itself, without any sediment layer, especially when not too coarse and not in too thin layers, has very great purifying powers, and, in addition, acts as a safeguard by positively preventing excessive rates of filtration on account of its frictional resistance. As an illustration take the case of a filter of sand with an effective size of 0.35 mm. and the minimum thickness of sand allowed by the German Board of Health, namely, one foot, and let us suppose that with clogging the loss of head has reached two feet to produce the desired velocity of 2.57 million gallons per acre daily. Suppose now that by some accident the sediment layer is suddenly broken or removed from a small area, the water will rush through this area, until a new sediment layer is formed, at a rate corresponding to the size, pressure, and depth of the sand, or 260 million gallons per acre daily—a hundred times the standard rate. Under these conditions the passing water will not be purified, but will pollute the entire effluent from the filter. Under corresponding conditions, with a deep filter of fine sand, say with an effective size of 0.20 mm. and 5 feet deep, the resulting rate would be only 17 million gallons per acre daily, or less than seven times the normal, and with the water passing through the full depth of fine sand, the resulting deterioration in the effluent before the sand again became so clogged as to reduce the rate to nearly the normal, would be hardly appreciable.

The results at Lawrence have shown that with very fine sands 0.09 and 0.14 mm., and 4 to 5 feet deep, with the quantity of water which can practically be made to pass through them, it is almost impossible to drive more than an insignificant fraction of the bacteria into the effluent. Even when the sands are entirely new, or have been scraped or disturbed in the most violent way, the first effluent passing, before the sediment layer could have been formed, is of good quality. Still finer materials, 0.04 to 0.06 mm., as far as could be determined, secured the absolute removal of all bacteria, but the rates of filtration which were possible were so low as to preclude their practical application.

With coarser sands, as long as the filter is kept at a steady rate of filtration, without interruptions of any kind, entirely satisfactory results are often obtained, although never quite so good as with the finer sands. Thus at Lawrence the percentages of bacteria (*B. prodigiosus*) appearing in the effluents under comparable conditions were as follows:

	1892	1893
With effective grain size 0.38 mm	....	0.16
" " " " 0.29 "	....	0.16
" " " " 0.26 "	....	0.10
" " " " 0.20 "	0.13	0.01
" " " " 0.14 "	0.04	0.03
" " " " 0.09 "	0.02	0.02

We may thus conclude that fine sands give normally somewhat better effluents than coarser ones, and that they are much more likely to give at least a tolerably good purification under unusual or improper conditions.

EFFECT OF GRAIN SIZE UPON FREQUENCY OF SCRAPING.

The practical objection to the use of fine sand is that it becomes rapidly clogged, so that filters require to be scraped at shorter intervals, and the sand washing is much more difficult and expensive. The quantities of water filtered between successive scrapings at Lawrence in millions of gallons per acre under comparable conditions have been as follows:

	1892	1893
Effective size of sand grain 0.38 mm	....	79
" " " " " 0.29 mm	....	70
" " " " " 0.26 mm	....	57
" " " " " 0.20 mm	58	....
" " " " " 0.14 mm	45	49
" " " " " 0.09 mm	24	14

The increase in the quantities passed between scrapings with increasing grain size is very marked.

With the fine sands, the depth to which the sand becomes dirty is much less than with the coarse sands, but as it is not generally practicable to remove a layer of sand less than about 0.6 inch thick, even when the actual clogged layer is thinner than this, the full quantity of sand has to be removed; and the quantities of sand to be removed and washed are inversely proportional to the quantities of water filtered between scrapings. On the other hand, with very coarse sands the sediment penetrates the sand to a greater depth than the 0.6 inch necessarily removed, so that a thicker layer of sand has to be removed, which may more than offset the longer interval. This happens occasionally in water-works, and a sand coarse enough to allow it occur is always disliked by superintendents, and is replaced with finer sand as soon as possible. It is obvious that the minimum expense for cleaning will be secured with a sand which just does not allow this deep penetration, and I am inclined to think that the sizes of the sands in use have actually been determined more often than otherwise in this way, and that the coarsest samples found, having effective sizes of about 0.40 mm., represent the practical limit to the coarseness of the sand, and that any increase above this size would be followed by increased expense for cleaning as well as by decreased efficiency.

SELECTION OF SAND.

In selecting a sand for filtration, when it is considered that repeated washings will remove some of the finest particles, and so increase slightly the effective size, a new sand coarser than 0.35 mm. would hardly be selected. Perhaps 0.20 might be given as a suitable lower limit. For comparatively clear lake- or reservoir-waters a finer sand could probably be used than would be the case with a turbid river-water. A mixed sand having a uniformity coefficient above 3.0 would be difficult to wash without separating it into portions of different sizes, and, in general, the lower the coefficient, that is, the more uniform the grain sizes, the better. Great pains should be taken to have the sand of the same quality throughout, especially in the same filter, as any variations in the grain sizes would lead to important variations in the velocity of filtration, the coarser sands passing more than their share of water (in proportion to the square of the effective sizes) and with reduced efficiency.

At Lawrence a sufficient quantity of natural sand was found of the grade required; but where suitable material cannot be so obtained it is necessary to use other methods. A mixed material can be screened from particles which are too large, and can be washed to free it from its finer portions, and in this way a good sand can be prepared, if necessary, from what might seem to be quite unpromising material. The methods of sand-washing will be described in Chapter V.

THICKNESS OF THE SAND LAYER.

The thickness of the sand layer is made so great that when it is repeatedly scraped in cleaning the sand will not become too thin for good filtration for a considerable time. When this occurs the removed sand must be replaced with clean sand. The original thickness of the sand in European filters is usually from 24 to 48 inches, thicknesses between 30 and 40 inches being extremely common, and this is reduced before refilling to from 12 to 24 inches. The Imperial Board of Health of Germany has fixed 12 inches as a limit below which the sand should never be scraped, and a higher limit is recommended wherever possible.

A thick sand layer has the same steadying action as a fine sand, and tends to prevent irregularities in the rate of filtration in proportion to its frictional resistance, and that without increasing the frequency of cleaning; but, on the other hand, it increases the necessary height of the filter, throughout, and consequently the cost of construction.

In addition to the steadying effect of a deep sand layer, some purification takes place in the lower part of the sand even with a good sediment layer on the surface, and the efficiency of deep filters is greater than that of shallow ones.

Layers of finer materials, as fine sand or loam, in the lower part of a filter, which would otherwise give increased efficiency without increasing the operating expenses, cannot be used. Their presence invariably gives rise sooner or later to sub-surface clogging at the point of junction with the coarser sand, as has been found by repeated tests at Lawrence as well as in some of the Dutch filters where such layers were tried; and as there is no object in putting a coarser sand under a finer, the filter sand is best all of the same size and quality from top to bottom.

UNDERDRAINING.

The underdrains of a filter are simply useful for collecting the filtered water; they play no part in the purification. One of the first requirements of successful filtration is that the rate of filtration shall be practically the same in all parts of the filter. This is most difficult to secure when the filter has just been cleaned and the friction of the sand layer is at a minimum. If the friction of the water in entering and passing through the underdrains is considerable, the more remote parts of the filters will work under less pressure, and will thus do less than their share of the work, while the parts near the outlet will be overtaxed, and filtering at too high rates will yield poor effluents.

To avoid this condition the underdrains must have such a capacity that their frictional resistance will be only a small fraction of the friction in the sand itself just after cleaning.

GRAVEL LAYERS.

The early filters contained an enormous quantity of gravel, but the quantity has been steadily reduced in successive plants.

Thus in 1866 Kirkwood, as a result of his observations, recommended the use of a layer four feet thick, and in addition a foot of coarse sand, while at the present time new filters rarely have more than two feet of gravel. Even this quantity seems quite superfluous, when calculations of its frictional resistance are made. Thus a layer of gravel with an effective size of 20 mm.* (which is much finer than that generally employed) only 6 inches thick will carry the effluent from a filter working at a rate of 2.57 million gallons per acre daily for a distance of 8 feet (that is, with underdrains 16 feet apart), with a loss of head of only 0.001 foot, and for longer distances tile drains are cheaper than gravel. To prevent the sand from sinking into the coarse gravel, intermediate sizes of gravel must be placed between, each grade being coarse enough so that there is no possibility of its sinking into the layer below. The necessary thickness of these intermediate layers is very small, the principal point being to have a layer of each grade at every point. Thus on the 6 inches of 20 mm. gravel mentioned above, three layers of two inches each, of 8 and 3 mm. gravel and coarse sand, with a total height of six inches, or other corresponding and convenient depths and sizes, would, if carefully placed, as effectually prevent the sinking of the filter sand into the coarse gravel as the much thicker layers used in the older plants.

The gravel around the drains should receive special attention. Larger stones can be here used with advantage, taking care that adequate spaces are left for the entrance of the water into the drains at a low velocity, and to make everything so solid in this neighborhood that there will be no chance for the stones to settle which might allow the sand to reach the drains.

At the Lawrence filter, at Königsberg in Prussia, at Amsterdam and other places, the quantity of gravel is reduced by putting the drains in trenches, so that the gravel is reduced from

---

* The method of calculating the size is given in Appendix III.

a maximum thickness at the drain to nothing half way between drains. The economy of the arrangement, however, as far as friction is concerned is not so great as would appear at first sight, and the cost of the bottom may be increased; but on the other hand it gives a greater depth of gravel for covering the drains with a small total amount of gravel.

As even a very small percentage of fine material is capable of getting in the narrow places and reducing the carrying power of the gravel, it is important that all such matters should be carefully removed by washing before putting the gravel in place. In England and Germany gravel is commonly screened for use in revolving cylinders of wire-cloth of the desired sizes, on which water is freely played from numerous jets, thus securing perfectly clean gravel. In getting gravel for the Lawrence filter, an apparatus was used, in which advantage was taken of the natural slope of the gravel bank to do the work, and the use of power was avoided. The respective grades of gravel obtained were even in size, and reasonably free from fine material, but it was deemed best to wash them with a hose before putting them in the filter.

To calculate the frictional resistance of water in passing gravel, we may assume that for the very low velocities which are actually found in filters the quantity of water passing varies directly with the head, which for these velocities is substantially correct, although it would not be true for higher rates, especially with the coarser gravels.* In the case of parallel underdrains the friction from the middle point between drains to the drains may be calculated by the formula:

$$\text{Total head} = \frac{1}{2} \frac{\text{Rate of filtration} \times (\frac{1}{2} \text{ distance between drains})^2}{\text{Average depth of gravel} \times \text{discharge coefficient}}.$$

The discharge coefficient for any gravel is 1000 times the quan-

---

* A full table of frictions with various velocities and gravels was given in the Rept. of Mass. State Board of Health, 1892, p. 555.

tity of water which will pass when $\frac{h}{l}$ is $\frac{1}{1000}$ expressed in million gallons per acre daily. The approximate values of this coefficient for different-sized gravels are as follows:

VALUES OF DISCHARGE COEFFICIENT.

For gravel with effective size 5 mm...... $c =$ 23,000
"   "   "   "   " 10 " ...... $c =$ 65,000
"   "   "   "   " 15 " ...... $c =$ 110,000
"   "   "   "   " 20 " ...... $c =$ 160,000
"   "   "   "   " 25 " ...... $c =$ 230,000
"   "   "   "   " 30 " ...... $c =$ 300,000
"   "   "   "   " 35 " ...... $c =$ 390,000
"   "   "   "   " 40 " ...... $c =$ 480,000

Example: What is the loss of head in the gravel at a rate of filtration of 2 million gallons per acre daily, with underdrains 20 feet apart, where the supporting gravel has an effective size of 35 millimeters, and is uniformly 1 ft. deep?

$$\text{Total head} = \frac{1}{2} \frac{2 \times 10^3}{1 \times 390{,}000} = .000256 \text{ ft.}$$

The total friction would be the same with the same average depth of gravel whether it was uniformly 1 foot deep, or decreasing from 1.5 at the drains to 0.5 in the middle, or from 2.0 to 0. The reverse case with the gravel layer thicker in the middle than at the drains does not occur and need not be discussed.

The depth of gravel likely to be adopted as a result of this calculation, when the drains are not too far apart, will be much less than that actually used in most European works, but as the two feet or more there employed are, I believe, simply the result of speculation, there is no reason for following the precedent where calculations show that a smaller quantity is adequate.

The reason for recommending a thin lower layer of coarse gravel, which alone is assumed to provide for the lateral movement of the water, is that if more than about six inches of gravel is required to give a satisfactory resistance, it will almost always be cheaper to use more drains instead of more gravel; and the reason for recommending thinner upper layers for preventing the sand from settling into the coarse gravel is that no failures of this portion of filters are on record, and in the few instances where really thin layers have been used the results have been entirely satisfactory. In Königsberg filters were built by Frühling,[*] in which the sand was supported by five layers of gravel of increasing sizes, respectively 1.2, 1.2, 1.6, 2.0, 3.2, or, together, 9.2 inches thick, below which there were an average of five inches of coarse gravel. These were examined after eight years of operation and found to be in perfect order.

At the Lawrence Experiment Station filters have been repeatedly constructed with a total depth of supporting gravel layers not exceeding six inches, and among the scores of such filters there has not been a single failure, and so far as they have been dug up there has never been found to have been any movement whatever of the sand into the gravel. The Lawrence city filter, built with corresponding layers, has shown no signs of being inadequately supported. In arranging the Lawrence gravel layers care has always been taken that no material should rest on another material more than three or four times as coarse as itself, and that each layer should be complete at every point, so that by no possibility could two layers of greater difference in size come together. And it is believed that if this is carefully attended to, no trouble need be anticipated, however thin the single layers may be.

UNDERDRAINS.

The most common arrangement, in other than very small filters, is to have a main drain through the middle of the filter,

---

[*] Frühling, Handbuch der Ingenieurwissenschaften, II. Band, VI. Kapitel.

with lateral drains at regular intervals from it to the sides. The sides of the main drain are of brick, laid with open joints to admit water freely, and the top is usually covered with stone slabs. The lateral drains may be built in the same way, but tile drains are also used and are cheaper. Care must be taken with the latter that ample openings are left for the admission of water at very low velocities. It is considered desirable to have these drains go no higher than the top of the coarsest gravel; and this will often control the depth of gravel used. If they go higher, the top must be made tight to prevent the entrance of the fine gravels or sand. Sometimes they are sunk in part or wholly (especially the main drain) below the floor of the filter. With gravel placed in waves, that is, thicker over the drains than elsewhere, as mentioned above, the drains are covered more easily than with an entirely horizontal arrangement. When this is done, the floor of the filter is trenched to meet the varying thickness of gravel, so that the top of the latter is level, and the sand has a uniform thickness.

Many filters (Lambeth, Brunswick, etc.) are built with a double bottom of brick, the upper layer of which, with open joints, supports the gravel and sand, and is itself supported by numerous small arches or other arrangements of brick, which serve to carry the water to the outlet without other drains. This arrangement allows the use of a minimum quantity of gravel, but is undoubtedly more expensive than the usual form, with only the necessary quantity of gravel; and I am unable to find that it has any corresponding advantages.

The frictional resistance of underdrains requires to be carefully calculated; and in doing this quite different standards must be followed from those usually employed in determining the sizes of water-pipes, as a total frictional resistance of only a few hundredths of a foot, including the velocity head, may cause serious irregularities in the rate of filtration in different parts of the filter.

The sizes of the underdrains differ very widely in proportion to the sizes of the filters in European works, some of them being excessively large, while in other cases they are so small as to suggest a doubt as to their allowing uniform rates of filtration, especially just after cleaning.

I would suggest the following rules as reasonably sure to lead to satisfactory results without making an altogether too lavish provision: In the absence of a definite determination to run filters at some other rate, calculate the drains for the German standard rate of a daily column of 2.40 meters, equal to 2.57 million gallons per acre daily. This will insure satisfactory work at all lower rates, and no difficulty on account of the capacity of the underdrains need be then anticipated if the rate is somewhat exceeded. The area for a certain distance from the main drain depending upon the gravel may be calculated as draining directly into it, provided there are suitable openings, and the rest of the area is supposed to drain to the nearest lateral drain.

In case the laterals are round-tile drains I would suggest the following limits to the areas which they should be allowed to drain:

Diameter of Drain.	To Drain an Area not Exceeding	Corresponding Velocity of Water in Drain.
4 inches	290 square feet.	0.30 foot.
6 "	750 " "	0.35 "
8 "	1530 " "	0.40 "
10 "	2780 " "	0.46 "
12 "	4400 " "	0.51 "

And for larger drains, including the main drains, their cross-sections at any point should be at least $\frac{1}{8000}$ of the area drained, giving a velocity of 0.55 foot per second with the rate of filtration mentioned above.

The total friction of the underdrains from the most remote points to the outlet will be friction in the gravel, plus friction in the lateral drains, plus the friction in main drain, plus the velocity head.

I have calculated in this way the friction of one of the Hamburg filters for the rate of 1,600,000 gallons per acre daily at which it is used. The friction was calculated for each section of the drains separately, so that the friction from intermediate points was also known. Kutter's formula was used throughout with $n = 0.013$. On the accompanying plan of the filter I have

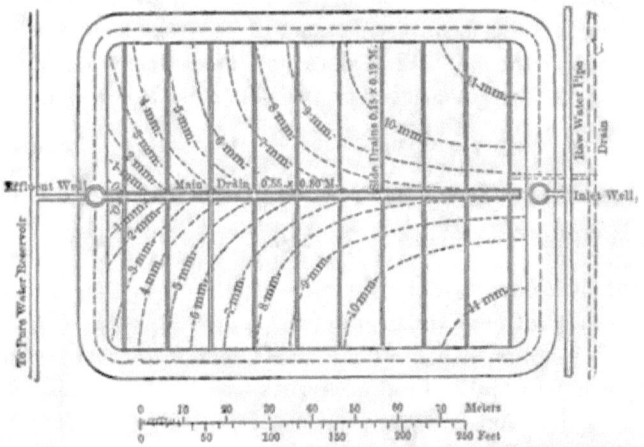

FIG. 4.—PLAN OF ONE OF THE HAMBURG FILTERS, SHOWING FRICTIONAL RESISTANCE OF THE UNDERDRAINS.

drawn the lines of equal frictional resistance from the junction of the main drain with the last laterals. My information was incomplete in regard to one or two points, so that the calculation may not be strictly accurate, but it is nearly so and will illustrate the principles involved.

The extreme friction of the underdrains is 11 millimeters = 0.036 foot.

The frictional resistance of the sand 39 inches thick, effective size 0.32 mm. and rate 1.60 million gallons per acre daily, when absolutely free from clogging, is by the formula, page 21, 15 mm., or .0490 foot, when the temperature is 50°. Practically there is some matter deposited upon the surface of the sand before filtration starts, and further, after the first scraping, there is some slight clogging in the sand below the layer removed by scraping. We can thus safely take the minimum frictional resistance of the sand including the surface layer at .07 foot. The average friction of the underdrains for all points is about .023 foot and the friction at starting will be .07 + .023 = .093 foot (including the friction in the last section to the effluent well where the head is measured, .100 foot, but the friction beyond the last lateral does not affect the uniformity of filtration). The actual head on the sand close to the outlet will be .093 and the rate of filtration $\frac{.093}{.070} \cdot 1.60 = 2.12$. The actual head at the most remote point will be .093 − .036 = .057, and the rate of filtration will there be $\frac{.057}{.070} \cdot 1.60 = 1.30$ million gallons per acre daily. The extreme rates of filtration are thus 2.12 and 1.30, instead of the average rate of 1.60. As can be seen from the diagram, only very small areas work at these extreme rates, the great bulk of the area working at rates much nearer the average. Actually the filter is started at a rate below 1.60, and the nearest portion never filters so rapidly as 2.12, for when the rate is increased to the standard, the sand has become so far clogged that the loss of head is more than the .07 foot assumed, and the differences in the rates are correspondingly reduced. Taking this into account, it would not seem that the irregularities in the rate of filtration are sufficient to affect seriously the action of the filter. They could evidently have been largely reduced by moderately increasing the sizes of the lower ends of the underdrains, where most of the friction occurs with the high velocities (up to .97 foot) which there result.

The underdrains of the Warsaw filters were designed by Lindley to have a maximum loss of head of only .0164 foot when filtering at a rate of 2.57, which gives a variation of only 10 per cent in the rates with the minimum loss of head of .169 foot in the entire filter assumed by him. The underdrains of the Berlin filters, according to my calculations, have .020 to .030 foot friction, of which an unusually large proportion is in the gravel, owing to the excessive distances, in some cases over 80 feet, which the gravel is required to carry the water. In this case, using less or finer gravel would obviously have been fatal, but the friction as well as the expense of construction would be much reduced by using more drains and less gravel.

The underdrains might appropriately be made slightly smaller, with a deep layer of fine sand, than under opposite conditions, as in this case the increased friction in the drains would be no greater in proportion to the increased friction in the sand itself.

The underdrains of a majority of European filters have water-tight pipes connecting with them at intervals, and going up through the sand and above the water, where they are open to the air. These pipes were intended to ventilate the underdrains and allow the escape of air when the filter is filled with water introduced from below. It may be said, however, that in case the drains are surrounded by gravel and there is an opportunity for the air to pass from the top of the drain into the gravel, it will so escape without special provision being made for it, and go up through the sand with the much larger quantity of air in the upper part of the gravel which is incapable of being removed by pipes connecting with the drains.

These ventilator pipes where they are used are a source of much trouble, as unfiltered water is apt to run down through cracks in the sand beside them, and, under bad management, unfiltered water may even go down through the pipes themselves. I am unable to find that they are necessary, except with underdrains so constructed that there is no other chance for the escape of air from the tops of them, or that they serve any useful purpose, while there are positive objections to their use. In some of the newer filters they have been omitted with satisfactory results.

DEPTH OF WATER ON THE FILTERS.

In the older works with but crude appliances for regulating the rate of filtration and admission of raw water, a considerable depth of water was necessary upon the filter to balance irregularities in the rates of filtration; the filter was made to be, to a certain extent, its own storage reservoir. When, however, appliances of the character to be described in Chapter IV are used for the regulation of the incoming water, and with a steady rate of filtration, this provision becomes quite superfluous.

With open filters a depth of water in excess of the thickness of any ice likely to be formed is required to prevent disturbance or freezing of the sand in winter. It is also frequently urged that with a deep water layer on the filter the water does not become so much heated in summer, but this point is not believed to be well taken, for in any given case the total amount of heat coming from the sun to a given area is constant, and the quantity of water heated in the whole day—that is, the amount filtered—is constant, and variations in the quantity exposed at one time will not affect the average resulting increase in temperature. If the same water remained upon the filter without change it would of course be true that a thin layer would be heated more than a deep one, but this is not the case.

It is also sometimes recommended that the depth of water should be sufficient to form a sediment layer before filtration starts, but this point would seem to be of doubtful value, especially where the filter is not allowed to stand a considerable time with the raw water upon it before starting filtration.

It is also customary to have a depth of water on the filter in excess of the maximum loss of head, so that there can never be a suction in the sand just below the sediment layer. It may be said in regard to this, however, that a suction below is just as effective in making the water pass the sand as an equal head above. At the Lawrence Experiment Station filters have been repeatedly used with a water depth of only from 6 to 12 inches, with losses of head reaching 6 feet, without the slightest inconvenience. The suction only commences to exist as the increasing head becomes greater than the depth of water, and there is no way in which air from outside can get in to relieve it. In these experimental filters in winter, when the water is completely saturated with air, a small part of the air comes out of the water just as it passes the sediment layer and gets into reduced pressure, and this air prevents the satisfactory operation of the filters. But this is believed to be due more to the warming and consequent supersaturation of the water in the comparatively warm places in which the filters stand than to the lack of pressure, and as not the slightest trouble is experienced at other seasons of the year, it may be questioned whether there would be any disadvantage at any time in a corresponding arrangement on a large scale where warming could not occur.

The depths of water actually used in European filters with the full depth of sand are usually from 36 to 52 inches. In only a very few unimportant cases is less than the above used, and only a few of the older works use a greater depth, which is not followed in any of the modern plants. As the sand becomes reduced in thickness by scraping, the depth of water is correspondingly increased above the figures given until the sand is replaced. The depth of water on the German covered filters is quite as great as upon corresponding open filters. Thus the Berlin covered filters have 51, while the new open filters at Hamburg have only 43 inches.

CHAPTER IV.

RATE OF FILTRATION AND LOSS OF HEAD.

THE rate of filtration recommended and used has been gradually reduced during the past thirty years. In 1866 Kirkwood found that 12 vertical feet per day, or 3.90 million gallons per acre daily, was recommended by the best engineers, and was commonly followed as an average rate. In 1868 the London filters averaged a yield of 2.18 million gallons* per acre daily, including areas temporarily out of use, while in 1885 the quantity had been reduced to 1.61. Since that time the rate has apparently been slightly increased.

The Berlin filters at Stralau constructed in 1874 were built to filter at a rate of 3.21 million gallons per acre daily. The first filters at Tegel were built for a corresponding rate, but have been used only at a rate of 2.57, while the more recent filters were calculated for this rate. The new Hamburg filters, 1892-3, were only intended to filter at a rate of 1.60 million gallons per acre daily. These in each case (except the London figures) are the standard rates for the filter-beds actually in service.

In practice the area of filters is larger than is calculated from these figures, as filters must be built to meet maximum instead of average daily consumptions, and a portion of the filtering area usually estimated at from 5 to 15 per cent, but in extreme cases reaching 50 per cent, is usually being cleaned, and so is for the time out of service. In some works also the rate of filtration on starting a filter is kept lower than the standard rate for a day or two, or the first portion of the effluent, supposed to be of inferior quality, is

---

* The American gallon is used throughout this book; the English gallon is one fifth larger.

wasted, the amount so lost reaching in an extreme case 9 to 14 per cent of the total quantity of water filtered.* In many of the older works also, there is not storage capacity enough for filtered water to balance the hourly fluctuations in consumption, and the filters must be large enough to meet the maximum hourly as well as the maximum daily requirements. For these reasons the actual quantity of water filtered in a year is only from 50 to 75 per cent of what would be the case if the entire area of the filters worked constantly at the full rate. A statement of the actual yields of a number of filter plants is given in Appendix IV. The figures for the average annual yields can be taken as quite reliable. The figures given for rate, in many cases, have little value, owing to the different ways in which they are calculated at different places. In addition most of the old works have no adequate means of determining what the rate at any particular time and for a single filter really is, and statements of average rates have only limited value. The filters at Hamburg are not allowed to filter faster than 1.60 or those at Berlin faster than 2.57 million gallons per acre daily, and adequate means are provided to secure this condition. Other German works aim to keep within the latter limit. Beyond this, unless detailed information in regard to methods is presented, statements of rate must be taken with some allowance.

EFFECT OF RATE UPON COST OF FILTRATION.

The size of the filters required, and consequently the first cost, depends upon the rate of filtration, but with increasing rates the cost is not reduced in the same proportion as the increase in rate, since the allowance for area out of use is sensibly the same for high and low rates, and in addition the operating expenses depend upon the quantity filtered and not upon the filtering area. Thus, to supply 10 million gallons at a maximum rate of 2 million gallons per acre daily we should require $10 \div 2 = 5$ acres $+$ 1 acre reserve for cleaning $= 6$ acres, while with a rate twice

---

* Piefke, *Zeitschrift für Hygiene*, 1894, p. 177.

as great, and with the same reserve (since the same amount of cleaning must be done, as will be shown below), we should require 10 ÷ 4 + 1 = 3.5 acres, or 58 per cent of the area required for the lower rate. Thus beyond a certain point increasing the rate does not effect a corresponding reduction in the first cost.

The operating cost for the same quantity of water filtered does not appear to be appreciably affected by the rate. It is obvious that at high rates filters will became clogged more rapidly, and will so require to be scraped oftener than at low rates, and it might naturally be supposed that the clogging would increase more rapidly than the rates, but this does not seem to be the case. At the Lawrence Experiment Station, under strictly parallel conditions and with identically the same water, filters running at various rates became clogged with a rapidity directly proportional to the rates, so that the quantities of water filtered between scrapings under any given conditions are the same whether the rate is high or low.

Of the eleven places (Appendix IV) in Germany filtering riverwaters from which statistics are available, four places with high rates, Lübeck, Stettin, Stuttgart, and Magdeburg, yielding 3.70 million gallons per acre daily, filtered on an average 59 million gallons per acre between scrapings. Three other places, Breslau, Altona, and Frankfurt, yielding 1.85, passed on an average 55 million gallons per acre between scrapings, and four other places, Bremen, Königsberg, Brunswick and Posen, yielding 1.34 million gallons per acre daily, passed only 40 million gallons per acre between scrapings. The works filtering at the highest rates thus filtered more water in proportion to the sand clogged than did those filtering more slowly, but I cannot think that this was the result of the rate. It is more likely that some of the places have clearer waters than others, and that this both allows the higher rate and causes less clogging than the more turbid waters.

A table of the estimated relative interest charges upon the costs of constructions, and of the annual operating expenses of filters at various rates will be given in Chapter IX.

EFFECT OF RATE UPON EFFICIENCY OF FILTRATION.

The effect of the rate of filtration upon the quality of the effluent has been repeatedly investigated. The efficiency almost uniformly decreases rapidly with increasing rate. Fränkel and Piefke * first found that with the high rates the number of bacteria passing some experimental filters was greatly increased. Piefke † afterward repeated these experiments, eliminating some of the features of the first series to which objection was made, and confirmed the first results. The results were so marked that Piefke was led to recommend the extremely low limit of 1.28 million gallons per acre daily as the safe maximum rate of filtration, but he has since repeatedly used 2.57 million gallons.

Kümmel,‡ on the other hand, in a somewhat limited series of experiments, was unable to find any marked connection between the rate and the efficiency, a rate of 2.57 giving slightly better results than rates of either 1.28 or 5.14.

The admirably executed experiments made at Zürich in 1886–8 upon this point, which gave throughout negative results, have but little value in this connection, owing to the extremely low number of bacteria in the original water.

At Lawrence in 1892 the following percentages of bacteria ($B.\ prodigiosus$) passed at the respective rates:

No. of Filter.	Depth.	Effective Size of Sand.	Rate. Million gallons per acre daily.				
			0.5	1.0	1.5	2.0	3.0
33A	60	0.14	0.002	......	......	0.040	......
34A	60	0.09	0.001	0.005	......	0.020	......
36A	60	0.20	......	......	0.050	......	0.050
37	60	0.20	......	......	0.010	0.130	......
38	24	0.20	0.018	......	0.140	0.110	0.310
39	12	0.20	0.014	0.070	......	0.080	0.520
40	12	0.20	......	0.070	......	0.090	......
42	12	0.20	0.016	......	......	0.150	0.550
Average			0.010	0.048	0.067	0.088	0.356

* *Zeitschrift für Hygiene*, 1891, page 38.
† *Journal für Gas- u. Wasserversorgung*, 1891, 208 and 228.
‡ *Journal für Gas- u. Wasserversorgung*, 1893, 161.

These results show a very marked decrease in efficiency with increasing rates, the number of bacteria passing increasing in general as rapidly as the square of the rate. The 1893 results also showed decreased efficiency with high rates, but the range in the rates under comparable conditions was less than in 1892, and the bacterial differences were less sharply marked.

While the average results at Lawrence, as well as most of the European experiments, show greatly decreased efficiency with high rates, there are many single cases, particularly with deep layers of not too coarse sand, where, as in Kümmel's experiments, there seems to be little connection between the rate and efficiency. An explanation of these apparently abnormal results will be given in Chapter VI.

It is commonly stated * that every water has its own special rate of filtration, which must be determined by local experiments, and that this rate may vary widely in different cases. Thus it is possible that the rate of 1.60 adopted at Hamburg for the turbid Elbe water, the rate of 2.57 used at Berlin, and about the same at London for much clearer river-waters, and the rate of 7.50 used at Zürich for the almost perfectly clear lake-water are in each case the most suitable for the respective waters. In other cases however, where rates much above 2.57 are used for river-waters, as at Lübeck and Stettin, there is a decided opinion that these rates are excessive, and in these instances steps are now being taken to so increase the filtering areas as to bring the rates within the limit of 2.57 million gallons per acre daily.

From the trend of European practice it would seem that for American river-waters the rate of filtration should not exceed 2.57 in place of the 3.90 million gallons per acre daily recommended by Kirkwood, or even that a somewhat lower rate might be desirable in some cases. Of course, in addition to the area

---

* Samuelson's translation of Kirkwood's "Filtration of River-waters;" Lindley, Die Nutzbarmachung des Flusswassers, *Journal für Gas- u. Wasserversorgung*, 1890, 501; Kaiserlichen Gesundheitsamt, Grundsätze für die Reinigung von Oberflachenwasser durch Sandfiltration; *Journal für Gas- u. Wasserversorgung*, 1894, Appendix I.

necessary to give this rate, a reserve for fluctuating rates and for cleaning should be provided, reducing the average yield to 2.00, 1.50, or even less. In the case of water from clear lakes, ponds, or storage reservoirs, especially when they are not subject to excessive sewage pollution or to strong algæ growths, it would seem that rates somewhat and perhaps in some cases very much higher (as at Zürich) could be satisfactorily used.

THE LOSS OF HEAD.

The loss of head is the difference between the heads of the waters above and below the sand layer, and represents the frictional resistance of that layer. When a filter is quite free from clogging this frictional resistance is small, but gradually increases with the deposit of a sediment layer from the water filtered until it becomes so great that the clogging must be removed by scraping before the process can be continued. After scraping the loss of head is reduced to, or nearly to, its original amount. With any given amount of clogging the loss of head is directly proportional to the rate of filtration; that is, if a filter partially clogged, filtering at a rate of 1.0, has a frictional resistance of 0.5 ft., the resistance will be doubled by increasing the rate to 2.00 million gallons per acre daily, provided no disturbance of the sediment layer is allowed. This law for the frictional resistance of water in sand alone also applies to the sediment layer, as I have found by repeated tests, although in so violent a change as that mentioned above, the utmost care is required to make the change gradually and prevent compression or breaking of the sediment layer. From this relation between the rate of filtration and the loss of head it is seen that the regulation of either involves the regulation of the other, and it is a matter of indifference which is directly and which indirectly controlled.

REGULATION OF THE RATE AND LOSS OF HEAD IN THE OLDER FILTERS.

In the older works, and in fact in all but a few of the newest works, the underdrains of the filters connect directly through a pipe with a single gate with the pure-water reservoir or pump-well, which is so built that the water in it may rise nearly or quite as high as that standing upon the filter.

A typical arrangement of this sort was used at the Stralau works at Berlin (now discontinued), Fig. 5. With this arrange-

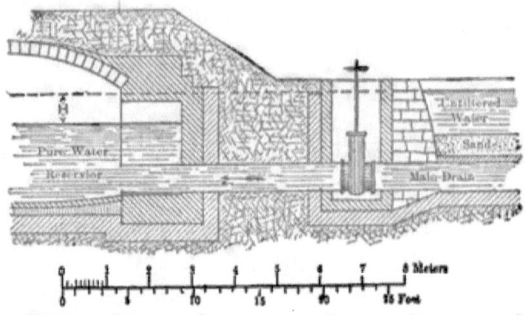

FIG. 5.—SIMPLEST FORM OF REGULATION: STRALAU FILTERS AT BERLIN.

ment the rate of filtration is dependent upon the height of water in the reservoir or pump-well, and so upon the varying consumption. When the water in the receptacle falls with increasing consumption the head is increased, and with it the rate of filtration, while, on the other hand, with decreasing draft and rising water in the reservoir, the rate of filtration decreases and would eventually be stopped if no water were used. This very simple arrangement thus automatically, within limits, adjusts the rate of filtration to the consumption, and at the same time always gives the highest possible level of water in the pump-well, thus also economizing the coal required for pumping.

In plants of this type the loss of head may be measured by floats on little reservoirs built for that purpose, connected with the underdrains; but more often there is no means of determining it, although the maximum loss of head at any time is the difference between the levels of the water on the filter and in the reservoir, or the outlet of the drain-pipe, in case the latter is above the water-line in the reservoir. The rate of filtration can only be measured with this arrangement by shutting off the incoming water for a definite interval, and observing the distance that the water on the filter sinks. The incoming water is regulated simply by a gate, which a workman opens or closes from time to time to hold the required height of water on the filter.

The only possible regulation of the rate and loss of head is effected by a partial closing of the gate on the outlet-pipe, by which the freshly-cleaned filters with nearly-closed gates are kept from filtering more rapidly than the clogged filters, the gates of which are opened wide. Often, however, this is not done, and then the fresh filters filter many times as rapidly as those which are partially clogged.

A majority of the filters now in use are built more or less upon this plan, including most of those in London and also the Altona works, which had such a favorable record with cholera in 1892.

The invention and application of methods of bacterial examination in the last years have led to different ideas of filtration from those which influenced the construction of the earlier plants. As a result it is now regarded as essential by most German engineers * that each filter shall be provided with devices for measuring accurately and at any time both the rate of filtration and the loss of head, and for controlling them, and also for making the rate independent of consumption by reservoirs for filtered water large enough to balance hourly variations (capacity $\frac{1}{4}$ to $\frac{1}{2}$ maximum daily quantity) and low enough so that they can never limit the rate of filtration by causing back-water on the filters. These points are now insisted upon by the German Imperial Board of Health,† and all new filters are built in accordance with them, while most of the old works are being built over to conform to the requirements.

---

* Lindley, *Journal für Gas- u. Wasserversorgung*, 1890, 501 ; Grahn, *Journal für Gas- u. Wasserversorgung*, 1890, 511 ; Halbertsma, *Journal für Gas. u. Wasserversorgung*, 1892, 686 ; Piefke, *Zeitschrift für Hygiene*, 1894, 151 ; and others.

† Appendix I.

APPARATUS FOR REGULATING THE RATE AND LOSS OF HEAD.

Many appliances have been invented for the regulation of the rate and loss of head. In the apparatus designed by Gill and used at both Tegel and Müggel at Berlin the regulation is effected by partially closing a gate through which the effluent passes into a chamber in which the water-level is practically constant (Fig. 6).

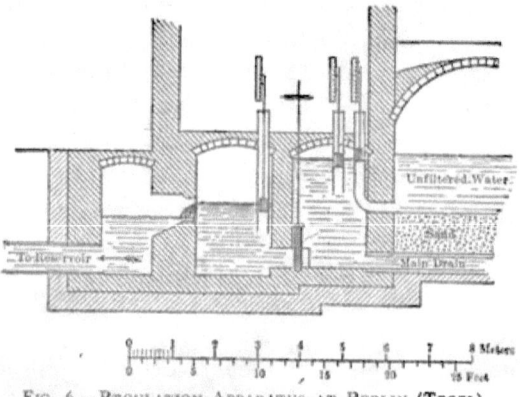

FIG. 6.—REGULATION APPARATUS AT BERLIN (TEGEL).

The rate is measured by the height of water on the weir which serves as the outlet for this second chamber into a third connecting with the main reservoir, while the loss of head is shown by the difference in height of floats upon water in the first chamber, representing the presure in the underdrains, and upon water in connection with the raw water on the filter. From the respective heights of the three floats the attendant can at any time see the rate of filtration and the loss of head, and when a change is required it is effected by moving the gate.

In the apparatus designed in 1866 by Kirkwood for St. Louis and never built (Fig. 7) the loss of head was directly, and the rate indirectly, regulated by a movable weir, which was to have been lowered from time to time by the attendant to secure the required results. This plan is especially remarkable as it meets the modern requirements of a regular rate independent of rate of consumption and of the water-level in the reservoir, and also allows continual measurements of both rate (height of water

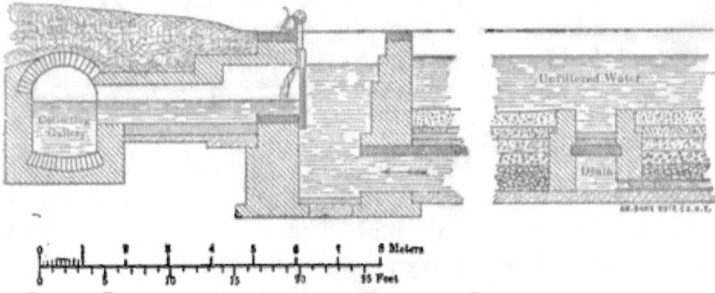

FIG. 7.—REGULATION APPARATUS AND SECTION OF FILTER RECOMMENDED FOR ST. LOUIS BY KIRKWOOD IN 1866.

on the weir) and head (difference in water-levels on filter and in effluent chamber) to be made, and control of the same by the position of the weir. Mr. Kirkwood found no filters in Europe with such appliances, and it was many years after his report was published before similar devices were used, but they are now regarded as essential.

The regulators for new filters at Hamburg (Fig. 8) are built

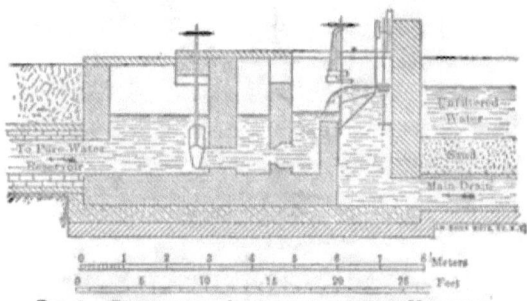

FIG. 8.—REGULATION APPARATUS USED AT HAMBURG.

upon the principle of Kirkwood's device, but provision is made for a second measurement of the water if desired by the loss of head in passing a submerged orifice. Both the rate and loss of head are indicated by a float on the first chamber connecting directly with the underdrain, which at the same time indicates the head on a fixed scale, the zero of which corresponds to the height of the water above the filter, and the rate upon a scale moving with the weir, the zero of which corresponds with the edge of the weir. The water on the filter is held at a perfectly constant level.

The regulators in use at Worms and those recently introduced at Magdeburg act upon the same principle, but the levels of the water on the filters are allowed to fluctuate, and the weirs and in fact, the whole regulating appliances are mounted on big floats in surrounding chambers of water connecting with the unfiltered water on the filters. I am unable to find any advantages in these appliances, and they are much more complicated than the forms shown by the cuts.

APPARATUS FOR REGULATING THE RATE DIRECTLY.

The above-mentioned regulators control directly the loss of head, and only indirectly the rate of filtration. The regulators

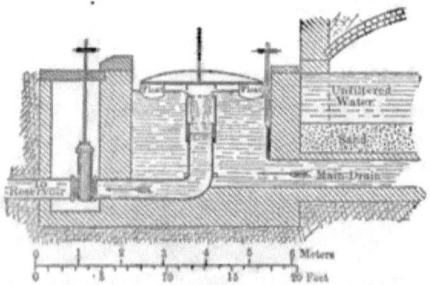

FIG. 9.—LINDLEY'S REGULATION APPARATUS AT WARSAW, RUSSIA.

at Warsaw were designed by Lindley to regulate the rate directly and make it independent of the loss of head. The quantity of water flowing away is regulated by a float upon the water in the effluent chamber, which holds the top of the telescope outlet-pipe a constant distance below the surface and so secures a constant rate. As the friction of the filter increases the float sinks with the water until it reaches bottom, when the filter must be scraped. A counter-weight reduces the weight on the float, and at the same time allows a change in the rate when desired. This apparatus is automatic. All of the other forms described require to be occasionally adjusted by the attendant, but the attention they require is very slight, and watchmen are always on duty at large plants, who can easily watch the regulators. The Warsaw apparatus is reported to work very satisfactorily, no trouble being experienced either by leaking or sticking of the telescope-joint, which is obviously the weakest point of the device, but fortunately a perfectly tight joint is not essential to the success of the apparatus. Regulators acting upon the same principle have recently been installed at Zürich, where they are operating successfully.

Burton[*] has described an ingenious device designed by him for the filters at Tokyo, Japan. It consists of a double acting valve of gun metal (similar to that shown by Fig. 11), through which the effluent must pass. This valve is opened and closed by a rod connecting with a piston in a cylinder, the opposite sides of which connect with the effluent pipe above and below a point where the latter is partially closed, so that the valve is opened and closed according as the loss of head in passing this obstruction is below or above the amount corresponding to the desired rate of filtration.

The use of the Venturi meter in connection with the regulation of filters would make an interesting study, and has, I believe, never been considered.

---

[*] The Water Supply of Towns. London, 1894.

APPARATUS FOR REGULATING THE HEIGHT OF WATER UPON FILTERS.

It will be seen by reference to the diagrams of the Berlin and Hamburg effluent regulators (Figs. 6 and 8) that their perfect operation is dependent upon the maintenance of a constant water-level upon the filters. The old-fashioned adjustment of the inlet-gate by the attendant is hardly accurate enough.

The first apparatus for accurately and automatically regulating the level of the water upon the filters was constructed at Leeuwarden, Holland, by the engineer, Mr. Halbertsma, who has since used a similar device at other places, and improved forms of which are now used at Berlin and at Hamburg.

At Berlin (Müggel) the water-level is regulated by a float upon the water in the filter which opens or shuts a balanced double valve on the inlet-pipe directly beneath, as shown in Fig. 10. It is not at all necessary that this valve should shut water-tight; it is only necessary that it should prevent the continuous inflow from becoming so great as to raise the water-level, and

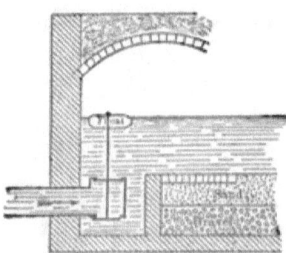

FIG. 10.—REGULATION OF INFLOW USED AT MÜGGEL, BERLIN.

for this reason loose, easily-working joints are employed. The apparatus is placed in a little pit next to the side of the filter, and the overflowing water is prevented from washing the sand by paving the sand around it for a few feet.

At Hamburg the same result is obtained by putting the valve

in a special chamber outside of the filter and connected with the float by a walking-beam (Fig. 11).

The various regulators require to be protected from cold and ice by special houses, except in the case of covered filters, where they can usually be arranged with advantage in the filter itself.

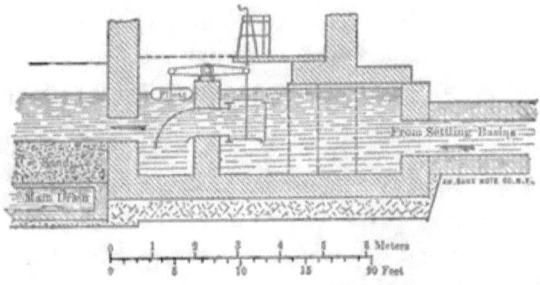

FIG. 11.—REGULATION OF INFLOW USED AT HAMBURG.

In regard to the choice of the form of regulator for both the inlets and outlets of filters, so far as I have been able to ascertain, each of the modern forms described as in use performs its functions satisfactorily, and in special cases any of them could properly be selected which would in the local conditions be the simplest in construction and operation.

LIMIT TO THE LOSS OF HEAD.

The extent to which the loss of head is allowed to go before filters are cleaned differs widely in the different works, some of the newer works limiting it sharply because it is believed that low bacterial efficiency results when the pressure is too great, although the frequency of cleaning and consequently the cost of operation are thereby increased.

At Darlington, England, I believe as a result of the German theories, the loss of head is limited to about 18 inches by a masonry weir built within the last few years. At Berlin, both at

Tegel and Müggel, the limit is 24 inches, while at the new Hamburg works 28 inches are allowed. At Stralau in 1893 an effort was made to not exceed a limit of 40 inches, but previously heads up to 60 inches were used, which corresponds with the 56 inches used at Altona; and, in the other old works, while exact information is not easily obtained because of imperfect records, I am convinced that heads of 60 or even 80 inches are not uncommon. At the Lawrence Experiment Station heads of 70 inches have generally been used, although some filters have been limited to 36 and 24 inches.

In 1866 Kirkwood became convinced that the loss of head should not go much above 30 inches, first, because high heads would, by bringing extra weight upon the sand, make it too compact, and, second, because when the pressure became too great the sediment layer on the surface of the sand, in which most of the loss of head occurs, would no longer be able to support the weight and, becoming broken, would allow the water to pour through the comparatively large resulting openings at greatly increased rates and with reduced efficiency.

In regard to the first point, a straight, even pressure many times that of the water on the filter is incapable of compressing the sand. It is much more the effect of the boots of the workmen when scraping that makes the sand compact. I have found sand in natural banks at Lawrence 70 or 80 feet below the surface, where it had been subjected to corresponding pressure for thousands of years, to be quite as porous as when packed in water in experimental filters in the usual way.

The second reason mentioned, or, as I may call it, the breaking-through theory, is very generally if not universally accepted by German engineers, and this is the reason for the low limit commonly adopted by them.

A careful study of the results at Lawrence fails to show the slightest deterioration of the effluents up to the limit used, 72 inches. Thus in 1892, taking only the results of the continuous filters of full height (Nos. 33A, 34A, 36A, and 37), we find that for the three days before scraping, when the head was nearly 72 inches, the averge number of bacteria in the effluents was 31 per cc., while for the three days after scraping, with very low heads, the number was 47. The corresponding numbers of *B. prodigiosus** were 1.1 and 2.7. This shows better work with the highest heads, but is open to the objection that the period just after scraping, owing to the disturbance of the surface, is commonly supposed to be a period of low efficiency.

To avoid this criticism in calculating the corresponding results for 1893, the numbers of the bacteria for the intermediate days which could not have been influenced either by scraping or by excessive head are put side by side with the others. Taking these results as before for continuous filters 72 inches high, and excluding those with extremely fine sands and a filter which was only in operation a short time toward the end of the year, we obtain the following results:

	Water Bacteria per cc.	B. Prodigiosus per cc.
Average 1st day after scraping, low heads............	79	6.1
" 2d " " " " "	44	4.1
" 3d " " " " "	45	3.6
Intermediate days, medium heads...................	59	4.5
Second from last day, heads of nearly 72 inches......	66	2.7
Next to the last day, " " " " "	56	3.2
Last day, " " " " "	83	2.5

These figures show a very slight increase of the water bacteria in the effluent as the head approaches the limit, but no such increase as might be expected from a breaking through of the sediment layer, and the *B. prodigiosus* which is believed to better indicate the removal of the bacteria of the original water,

---

* A special species of bacteria artificially added to secure more precise information n regard to the passage of germs through the filter.

actually shows a decrease, the last day being the best day of the whole period.

The Lawrence results, then, uniformly and clearly point to a conclusion directly opposite to the commonly accepted view, and I have thus been led to examine somewhat closely the grounds upon which the breaking-through theory rests.

The two works which have perhaps contributed most to the theories of filtration are the Stralau and Altona works. After examining the available records of these works, I am quite convinced that at these places there has been, at times at least, decreased efficiency with high heads. For the Stralau works this is well shown by Piefke's plates in the *Zeitschrift für Hygiene*, 1894, after page 188. In both of these works, however, the apparatus (or lack of apparatus) for regulating the rate is that shown by Fig. 5, page 49, and the rate of filtration is thus dependent upon the rate of consumption and the height of water in the reservoir. At the Stralau works, at the time covered by the above-mentioned diagrams, the daily quantity of water filtered was 27 times the capacity of the reservoir, and the rate of filtration must consequently have adapted itself to the hourly consumptions. The data which formed the basis of Kirkwood's conclusions are not given in detail, but it is quite safe to assume that they were obtained from filters regulated as those at Altona and Stralau are regulated, and what is said in regard to the latter will apply equally to his results.

Piefke[*] shows that among the separate filters at Stralau, all connected with the same pure-water reservoir, those connected through the shorter pipes gave poorer effluents than the more remote filters, and he attributes the difference to the frictional resistance of the connecting pipes, which helped to prevent excessive rates in the filters farthest away when the water in the reservoir became low, and thus the fluctuations in the rates in these filters were less than in those close to the reservoir. He

---

[*] *Zeitschrift für Hygiene*, 1894, p. 173.

does not, however, notice, in speaking of the filters in which the decreased efficiencies with high heads were specially marked, that they follow in nearly the same order, and that of the four open filters mentioned three were near the reservoir and only one was separated by a comparatively long pipe, indicating that the deterioration with high heads was only noticeable, or at least was much more conspicuous, in those filters where the rates fluctuated most violently.

It requires no elaborate calculation to show that of two filters connected with the same pure-water reservoir, as shown by Fig. 5, with only simple gates on the connecting pipes, one of them clean and throttled by a nearly closed gate, so that the normal pressure behind the gate is above the highest level of water in the reservoir, and the other clogged so that the normal pressure of the water in the drain is considerably below the highest level of the water in the reservoir, the latter will suffer much the more severe shocks with fluctuating water-levels; and the fact being admitted that fluctuating levels are unfavorable, we must go farther and conclude that the detrimental action will increase with increasing loss of head. I am inclined to think that this theory is adequate to explain the Stralau and Altona results without resource to the breaking-through theory.

While the above does not at all prove the breaking-through theory to be false, it explains the results upon which it rests in another way, and can hardly fail to throw so much doubt upon it as to make us refuse to allow its application to those works where a regular rate of filtration is maintained regardless of variations in the consumption, until proof is furnished that it is applicable to them.

I have been totally unable to find satisfactory European results in regard to this point. The English works can furnish nothing, both on account of the lack of regulating appliances and because the monthly bacterial examinations are inadequate for a discussion of hourly or daily changes. The results from the older Continental works are also excluded for one or the other, or more often for both, of the above reasons. The Hamburg, Tegel, and Müggel results, so far as they go, show no deterioration with increased heads, but the heads are limited to 24 or 28 inches by the construction of the filters, and the results thus entirely fail to show what would be obtained with heads more than twice as high.

I am thus forced to conclude that there is no adequate evidence of inferior efficiency with high heads in filters where the rates are independent of the water-level in the pure-water reservoir, the only results directly to the point—the Lawrence results mentioned above—indicating that the full efficiency is maintained with heads reaching at least 72 inches.

The principal reason for desiring to allow a considerable loss of head is an economical one; the period will then be lengthened, while the frequency of scraping and the volume of sand to be washed and replaced will be correspondingly reduced. There may be other advantages in long periods, such as less trouble with scraping and better work in cold winter weather, but the cost is the most important consideration.

It is the prevalent idea among the German engineers that the loss of head after reaching 24 to 30 inches would increase very rapidly, so that the quantity of water filtered, in case a much higher head was allowed, would not be materially increased. No careful investigations, however, have been made, and indeed they are hardly possible with existing arrangements, as in the older filters the loss of head fluctuates with varying rates of filtration in such a way that only results of very doubtful value can be obtained, and in the newer works the loss of head is too closely limited, and the curves which can be drawn by extrapolation are evidently no safe indications of what would actually happen if the process was carried farther.

On the other hand, I was told by the attendant at Darlington, England, that since the building of the weir a few years ago, which now limits the loss of head to about 18 inches **instead of the 5 feet or more formerly used,** the quantity of sand to be removed has been three times **as great** as formerly. No records are kept, and this can only be given as the general impression of the man who superintends the work.

At Lawrence the average quatities of water filtered between scrapings with sand of an effective size **of 0.20 mm.** have been **as** follows:

Maximum **Loss of** Head.	Million Gallons per **Acre filtered** between Scrapings.		
	1892.	1893.	Average.
70 inches	58	88	73
34 "	32	22	27
22 "	17	16	16

With sand of an **effective size of 0.29 mm.** the results were:

	1893.
70 inches	70
22 "	29

These results **indicate a great increase in the quantity of water** filtered between **scrapings with increasing heads,** the figures being nearly proportional to the maximum **heads** used in the respective cases. It is, of course, quite possible that the results would differ in different places with **the** character of the raw water and of the filtering material.

The depth of sand to be removed by scraping at one time is, **within** limits, practically independent of the quantity **of dirt** which it has accumulated, and any lengthening of the period means **a** corresponding reduction in the quantity of sand to be removed, washed **and** replaced and consequently an important **reduction in the operating cost, as well as a** reduction in the **area of filters out of use while** being cleaned, and so, in the capital **cost.**

Among the minor objections to an increased loss of head are the greater **head** against which **the** water must be pumped, and the possible increased difficulty of filling filters with filtered water from below after scraping, but these would hardly have much weight against the economy indicated by the Lawrence experiments for the higher heads.

High heads will also drive an increased quantity of water through any cracks or passages in the filter. Such leaks have at last been found to be the cause of the inferior work of the covered filters at Stralau, the water going down unfiltered in certain corners, especially at high heads; but with careful construction there should be no cracks, and with the aid of bacteriology to find the possible leaks this ought not to be a valid objection.

In conclusion: the trend of opinion is strongly in favor of limiting the loss of head to about 24 to 30 inches as was suggested by Kirkwood, but I am forced to conclude that there is reason to believe that equally good results can be obtained with lower operating expenses by allowing higher heads to be used, at least in the case of filters with modern regulating appliances, and, I would suggest that filters should be built so as not to exclude the use of moderately high heads, and that the limit to be permanently used should be determined by actual tests of efficiency and length of period with various losses of head after starting the works.

CHAPTER V.

CLEANING FILTERS.

When a filter has become so far clogged that it will no longer pass a satisfactory quantity of water with the allowable head it must be cleaned by scraping off and removing the upper layer of dirty sand.

To do this without unnecessary loss of time the unfiltered water standing upon the filter is removed by a drain above the sand provided for that purpose. The water in the sand must then be lowered below the surface of the sand by drawing water from the underdrains until the sand is firm enough to bear the weight of the workmen. By the time that this is accomplished the last water on the surface should have soaked away, and the filter is ready to be scraped. This is done by workmen with wide, sharp shovels, and the sand removed is taken to the sand-washing apparatus to be washed and used again. Special pains are given to securing rapid and cheap transportation of the sand. In some cases it is wheeled out of the filter on an inclined plane to the washer. In other cases a movable crane is provided which lifts the sand in special receptacles and allows it to fall into cars on a tram-line on which the crane also moves. The cars as filled are run to the washer and also serve to bring back the washed sand. When the dirty sand has been removed, the surface of the sand is carefully smoothed and raked. This is especially necessary to remove the effects of the workmen's boots.

It is customary in the most carefully managed works to fill the sand with filtered water from below, introduced through the underdrains. In case the ordinary level of the water in the pure-water canal is higher than the surface of the sand in the filters, this is accomplished by simply opening a gate provided for the purpose, which allows the water to pass around the regulating apparatus. Otherwise filters can be filled from a special pipe taking its water from any filter which at that time can deliver its effluent high enough for that purpose. The quantity of water required for filling the sand from below is ordinarily but a fraction of one per cent of the quantity filtered.

Formerly, instead of filling from below, after cleaning, the raw water was brought directly onto the surface of the filter. This was said to only imperfectly fill the sand-pores, which still contained much air. If, however, the water is not brought on too rapidly it will sink into the sand near the point where it is applied, pass laterally through the sand or underlying gravel to other parts of the filter, and then rise, so that even in this case all but a little of the filter will be really filled from below. This is, however, open to the objection that however slowly the water is introduced, the sand which absorbs it around the inlet filters it at a very high rate and presumably imperfectly, so that the water in the underdrains at the start will be poor quality and the sand around the inlet will be unduly clogged. The practice of filling from below is therefore well founded.

As soon as the surface of the sand is covered with the water from below, raw water is introduced from above, filling the filter to the standard height, care being taken at first that no currents are produced which might wash the surface of the sand. It has been recommended by Piefke and others that this water should be allowed to stand for a time up to twenty-four hours before starting the filtration, to allow the formation of a sediment layer, and in some places, especially at Berlin and the works of some of the London companies, this is done; but varying importance is attached to the procedure, and it is invariably omitted, so far as I can learn, when the demand for water is heavy.

The depth of sand removed by scraping must at least equal

the depth of the discolored layer, but there is no sharp dividing line, the impurities gradually decreasing from the surface downward. Fig. 12 shows the relative number of bacteria found in the sand at various depths in one of the Lawrence experimental filters, and is a representative result, although the actual numbers vary at different times. In general it may be said that the bulk of the sediment is retained in the upper quarter inch, but it is desirable to remove also the less dirty sand below and, in fact, it is apparently impossible with the method of scraping in use to remove so thin a layer as one fourth inch. Practically

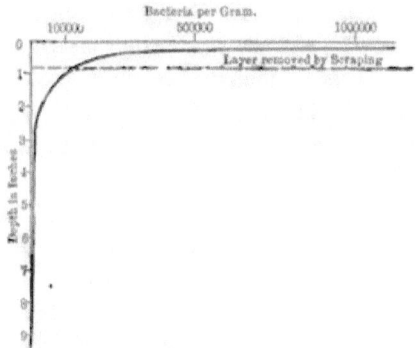

FIG. 12.—DIAGRAM SHOWING ACCUMULATION OF BACTERIA NEAR THE SURFACE OF THE SAND.

the depth to which sand is removed is stated to be from 0.40 to 1.20 inch. Exact statistics are not easily obtained, but I think that 2 centimeters or 0.79 inch may be safely taken as about the average depth usually removed in European filters, and it is this depth which is indicated on Fig. 12.

At the Lawrence Experiment Station, the depth removed is often much less than this, and depends upon the size of grain of the sand employed, the coarser sands requiring to be more deeply scraped than the finer ones. The method of scraping, however, which allows the removal of very thin sand layers, is only possible because of the small size of the filters, and as it is incapable of application on a large scale, the depths thus removed are only interesting as showing the results which might be obtained in practice with a more perfect method of scraping.

The replacing of the washed sand is usually delayed until the filter has been scraped quite a number of times—commonly for a year. The last scraping before refilling is much deeper than usual, because the sand below the depth of the ordinary scraping is somewhat dirty, and might cause trouble if left below the clean sand.

In England it is the usual if not the universal practice to replace the washed sand at the bottom between the old sand and the gravel. This is done by digging up the entire filter in sections about six feet wide. The old sand in the first section is removed clear down to the gravel, and the depth of washed sand which is to be replaced is put in its place. The old sand from the next six-foot section is then shovelled upon the first section of clean sand, and its place is in turn filled with fresh sand. With this practice the workmen's boots are likely to disturb the gravel each year, necessitating a thicker layer of the upper and finest grade than would otherwise be required.

In Germany this is also sometimes done, but more frequently the upper layer of slightly clogged sand below the regular scraping is removed as far as the slightest discoloration can be seen, perhaps 6 inches deep. The sand below is loosened for another 6 inches and allowed to stand dry, if possible, for some days; afterwards the washed sand is brought on and placed above. The washed sand is never replaced without some such treatment, because the slightly clogged sand below the layer removed would act as if finer than the freshly washed sand,* and there would be a tendency to sub-surface clogging.

---

* Report Mass. State Board of Health for 1891, p. 438; 1892, page 409.

FREQUENCY OF SCRAPING.

The frequency of scraping depends upon the character of the raw water, the thoroughness of the preliminary sedimentation, the grain-size of the filter sand, the rate of filtration, and the maximum loss of head allowed. With suitable conditions the period between scrapings should never be less than one week, and will but rarely exceed two months. Under exceptional conditions, however, periods have been recorded as low as one day and as high as one hundred and ten days. Periods of less than a week's duration are almost conclusive evidence that something is radically wrong, and the periods of one day mentioned were actually accompanied by very inadequate filtration. In 1892 the average periods at the German works varied from 9.5 days at Stettin (with an excessive rate) to 40 days at Brunswick, the average of all being 25 days.*

The quantity of water per acre filtered between scrapings forms the most convenient basis for calculation. The effect of rate (page 45), loss of head (page 61), and size of sand grain (page 28) have already been discussed, and it will suffice to say here that the total quantity filtered between scrapings is apparently independent of the rate of filtration, but varies with the maximum loss of head and with the grain-size of the sand, and apparently nearly in proportion to them. Eleven German filter-works in 1892, drawing their waters from rivers, filtered on an average 51 million gallons of water per acre between scrapings, the single results ranging from 28 at Bremen to 71 at Stuttgart, while Zürich, drawing its water from a lake which is but very rarely turbid, filtered 260 million gallons per acre between scrapings. Unfortunately, the quantities at Berlin, where (in 1892 two thirds and now all) the water is drawn from comparatively large ponds on the rivers, are not available for comparison.

At London, in 1884, the average quantities of water filtered

---
* Appendix IV.

between scrapings varied from 43 to 136 million gallons per acre with the different companies, averaging 85, and in 1892 the quantities ranged from 73 to 157, averaging 90 million gallons per acre. The greater quantity filtered at London may be due to the greater sizes of the sedimentation-basins, which for all the companies together hold a nine days' supply at London against probably less than one day's supply for the German works.

There is little information available in regard to the frequency of scraping with water drawn from impounding reservoirs. In some experiments made by Mr. FitzGerald at the Chestnut Hill reservoir, Boston, the results of which are as yet unpublished, a filter with sand of an effective size of only .09 mm. averaged 58 million gallons per acre between scrapings for nine periods, the rate of filtration being 1.50 million gallons per acre daily, while another filter, with sand of an effective size of .18 mm., passed an average of 93 million gallons per acre for ten periods at the same rate. These experiments extended through all seasons of the year, and taking into account the comparative fineness of the sands they show rather high quantities of water filtered between scrapings.

The quantity of water filtered between scrapings is usually greatest in winter, owing to the smaller quantity of sediment in the raw water at this season, and is lowest in times of flood, regardless of season. In summer the quantity is often reduced to a very low figure in waters supporting algæ growths, especially when the filters are not covered. Thus at Stralau in 1893 during the algæ period the quantity was reduced to 14 million gallons per acre for open filters,* but this was quite exceptional, the much-polluted, though comparatively clear, Spree water furnishing unusually favorable conditions for the algæ.

---

* Piefke, *Zeitschrift für Hygiene*, 1894, p. 177.

QUANTITY OF SAND TO BE REMOVED.

In regard to the quantity of sand to be removed and washed, if we take the average result given above for the German works filtering river-waters of 51,000,000 gallons per acre filtered between scrapings, and the depth of sand removed at two centimeters or 0.79 inch, we find that one volume of sand is required for every 2375 volumes of water filtered, or 2.10 cubic yards per million gallons. At Bremen, the highest average result, the quantity would be 3.80 yards, and at Stralau during the algæ season 7.70 yards. At Zürich, on the other hand, the quantity is only 0.41 yard, and at London, with 87,000,000 gallons per acre filtered between scrapings, the quantity of sand washed would be 1.24 yards per million gallons; assuming always that the layer removed is 0.79 inch thick.

These estimates are for the regular scrapings only, and do not include the annual deeper scraping before replacing the sand, which would increase them by about one third.

WASTING THE EFFLUENTS AFTER SCRAPING.

It has already been stated that an important part of the filtration takes place in the sediment layer deposited on top of the sand from the water. When this layer is removed by scraping its influence is temporarily removed, and reduced efficiency of filtration may result. The significance of this reduced efficiency became apparent when the bacteria in the water were studied in their relations to disease, and Piefke suggested * that the first effluent after scraping should be rejected for one day after ordinary scrapings and for one week after replacing the sand. In a more recent paper † he reduces these estimates to the first million gallons of water per acre filtered after scraping

---

* *Journal für Gas- und Wasserversorgung*, 1887, p. 595.
† *Zeitschrift für Hygiene*, 1894, p. 172.

for open and twice as great a quantity for covered filters, and to six days after replacing the sand, which last he estimates will occur only once a year. Taking the quantity of water filtered between scrapings at 13.9 million gallons per acre, the quantity observed at Stralau in the summer of 1893, he finds that it is necessary to waste 9 per cent of the total quantity of effluent from open and 13.8 per cent of that from covered filters.

The eleven German water-works * filtering river-waters, however, filtered on an average 51.0 instead of 13.9 million gallons per acre between scrapings, and applying Piefke's figures to them the quantities of water to be wasted would be only about one fourth of his estimates for Stralau.

The rules of the Imperial Board of Health † require that every German filter shall be so constructed "that when an inferior effluent results it can be disconnected from the pure-water pipes and the filtrate allowed to be wasted." The drain-pipe for removing the rejected water should be connected below the apparatus for regulating the rate and loss of head, so that the filter can be operated exactly as usual, and the effluent can be turned back to the pure-water pipes without stopping or changing the rate. The works at Berlin and at Hamburg conform to this requirment, and most of the older German works have been or are being built over to make them do so.

In regard to the extent of deterioration after scraping, Piefke's experiments have always shown much larger numbers of bacteria both of the ordinary forms and of special applied forms on the first day after scraping, the numbers frequently being many times as high as at other times.

At the Lawrence Experiment Station it was found in 1892 that on an average the number of water bacteria was increased by 70 per cent (continuous filters only) for the three days following scraping, while *B. prodigiosus* when applied was increased 140 per

---

* Appendix IV. † Appendix I.

cent, the increase being most **marked** where the depth of sand was **least**, and with the highest rate of filtration.

The same tendency was found in 1893, when the increase in the water bacteria on the first **day after** scraping was only 19 per cent and *B. prodigiosus* 64 per cent, but for a portion of the year the difference was greater, averaging 132 and 262 per cent, respectively. These differences **are much less** than those recorded by Piefke, and with the high efficiencies regularly obtained **at** Lawrence they would **hardly justify** the expensive practice of wasting the effluent.

The reduction in efficiency following scraping **is much** less at low **rates, and if a** filter **is started** at much **less than** its normal **rate** after scraping, and then gradually increased to the standard after the sediment layer is formed, the poor work will **be** largely **avoided.** Practically **this is** done at Berlin and at Hamburg. **The filters are** started **at a fourth or** less of the usual rates and **are** gradually increased, as past experience with bacterial results **has shown it** can be safely done, and the effluent is then even at first so **well** purified that it need not be wasted.

Practically in building **new** filters the provision of a suitable connection for wasting the **effluents into the** drain which is necessary for emptying **them involves** no serious expense and should **be** provided, **but it may be** questioned **how often it should be used** for wasting the effluents. **If the raw** water is **so** bad that a good effluent cannot be obtained by careful manipulation even just after scraping, the course of the Berlin authorities in closing the Stralau works and seeking a less polluted supply would **seem to be** the only really safe procedure.

SAND-WASHING.

The sand-washing apparatus **is an** important part of most European filtering plants. It seldom happens that a natural sand can be **found clean** enough and sufficiently free from fine particles, although such a sand was found and used for the Lawrence filter. Most of the sand in use for filtration in Europe was originally washed. In the operation of the filters also, sand-washing is used for the dirty sand, which can then be used over and over at a much lower cost than would be the case if fresh sand was used for refilling. The methods used for washing sand at the different works present a great variety both in their details and in the underlying principles. Formerly boxes with double perforated bottoms in which the sand was placed and stirred by a man as water from below rose through them, and other similar arrangements were commonly used, but they are at present only retained, so far as I know, in some of the smaller English works. The cleansing obtained is apparently considerably less thorough than with some of the modern devices.

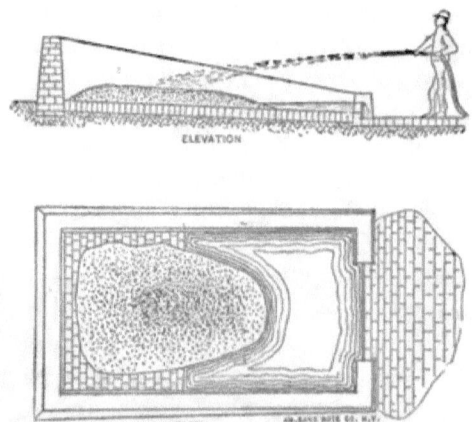

FIG. 13.—HOSE-WASHING FOR DIRTY SAND.

Hose-washing is used in London by the Southwark and Vauxhall, Lambeth and Chelsea companies, and also at Antwerp. For this a platform is constructed about 15 feet long by 8 feet wide, with a pitch lengthwise of 6 to 8 inches (Fig. 13). The platform is surrounded by a wall rising from one foot at the bottom to three feet high at the top, except the lower end, which is closed by a removable plank weir 5 or 6 inches high. From two to four cubic yards of the sand are placed upon this platform and a stream of water from a hose with a $\frac{3}{4}$ or $\frac{7}{8}$-inch nozzle is played upon it, moving it about from place to place. The sand itself is always kept toward the upper end of the platform, while the water with the dirt removed flows down into the pond made by the weir, where the sand settles out and the dirt overflows with the water. When the water comes off clear, which is usually after an hour or a little less, the weir is removed, and, after draining, the sand is removed. These arrangements are built in pairs so that the hose can be used in one while the sand is being changed in the other. They are usually built of brick laid in cement, but plank and iron are also used. The corners are sometimes carried out square as in the figure, but are more often rounded. The washing is apparently fairly well done.

In Germany the so-called "drum" washing-machine, drawings of which have been several times published,* has come to be almost universally used. It consists of a large revolving cylinder, on the bottom of the inside of which the sand is slowly pushed up toward the higher end by endless screw-blades attached to the cylinder, while water is freely played upon it all the way. The machine requires a special house for its accommodation and from 2 to 4 horse-power for its operation. It washes from 2.5 to 4 yards of sand per hour most thoroughly, with a consumption of from 11 to 14 times as large a volume of water. The apparatus is not patented or made for sale, but full plans can be easily secured.

A machine made by Samuel Pegg & Sons, Leicester, Eng., pushes the sand up a slight incline down which water flows. It is very heavy and requires power to operate it. The patent has

---

* *Glaser's Annalen*, 1886, p. 48; *Zeit. f. Hygiene*, 1889, p. 128.

expired. A machine much like it but lighter and more convenient and moved by water-power derived from the water used for washing instead of steam-power is used at Zürich with good results.

In Greenway's machine the sand is forced by a screw through a long narrow cylinder in which there is a current of water in the opposite direction. The power required is furnished by a water-motor, as with the machine at Zürich. The apparatus is mounted on wheels and is portable; it has an appliance for piling up the washed sand or loading it onto cars. It is patented and is manufactured by James Gibb & Co., London.

Several of the London water companies are now using ejector washers, and such an apparatus has been placed by the side of the "drum" washers at Hamburg. This apparatus was made by Körting Brothers in Hannover, and combines the ejectors long made by that firm with hoppers from designs by Mr. Bryan, engineer of the East London Water Company. An apparatus differing from this only in the shape of the ejectors and some minor details has been patented in England, and is for sale by Messrs. Hunter, Frazer & Goodman, Bow, London.

Both of these forms consist of a series of conical hoppers, from the bottom of each of which the sand and water are forced into the top of the next by means of ejectors, the excess of dirty water overflowing from the top of each hopper. The apparatus is compact and not likely to get out of order, but is not portable. It can be easily arranged to take the sand at the level of the ground, or even lower if desired, and deliver it washed at some little elevation, thus minimizing hand-labor. The washing is regular and thorough. The objection most frequently raised against its use is the quantity of water required, but at Hamburg I was informed that the volume of water required was only about 15 times that of the sand, while almost as much (13-14 volumes) were required for the "drum" washers, and the saving in power much more than offset the extra cost for water.

In addition to the above processes of sand-washing, Piefke's method of cleaning without scraping * might be mentioned, although as yet it has hardly passed the experimental stage, and has only been used on extremely small filters. The process consists of stirring the surface sand of the filter with "waltzers" while a thin sheet of water rapidly flows over the surface. This arrangement necessitates a special construction of the filters, providing for rapidly removing the unfiltered water from the surface, and for producing a regular and rapid movement of a thin sheet of water over the surface. In the little filters now in use, one of which I saw in a brewery in Berlin, the cleaning is rapidly, cheaply, and apparently well done.

In washing dirty sand it is obvious that any small sand-grains will be removed with the dirt, and in washing new sand the main object is to remove the grains below a certain size. It is also apparent that the sizes of grains which will and those which will not be removed are dependent upon the mechanical arrangements of the washer, as, for example, with the ejectors, upon the sizes of the hoppers, and the quantity of water passing through them, and care should be taken to make them correspond with the size of grain selected for the filter sand. This can only be done by experiment, as no results are available on this point.

In some places filtered water is used for sand-washing, although this seems quite unnecessary, as ordinary river-water answers very well. It is, however, often cheaper, especially in small works, to use the filtered water from the mains rather than provide a separate supply for the washers.

The quantity of water required for washing may be estimated at 15 times the volume of the sand and the sand as 0.04 per cent of the volume of the water filtered (page 70), so that

---

* *Vierteljahresschrift für öffentliche Gesundheitspflege,* 1891, p. 59.

0.6 per cent of the total quantity of water filtered will be required for sand-washing.

The cost of sand-washing in Germany with the "drum" washers is said to be from 14 to 20 cents per cubic yard, including labor, power, and water. In America the water would cost no more, but the labor would be perhaps twice as dear. With an ejector apparatus I should estimate the cost of washing dirty sand as follows: The sand would be brought and dumped near to the washer, and one man could easily feed it in, as no lifting is required. Two men would probably be required to shovel the washed sand into barrows or carts with the present arrangements, but I think with a little ingenuity this handling could be made easier.

ESTIMATED COST OF OPERATING EJECTOR WASHERS 9 HOURS.

Wages of 3 men at $2.00.............................. $6.00
110,000 gals. water (15 times the volume of sand)
  at 0.05 a thousand gals................................ 5.50

Total cost of washing 36 cubic yards............ $11.50
  or 32 cents a cubic yard.

The cost of washing new sand might be somewhat less. The other costs of cleaning filters, scraping, transporting, and replacing the sand are much greater than the washing itself. Lindley states that at Warsaw 29 days' labor of 10 hours for one man are required to scrape an acre of filter surface, and four times as much for the annual deep scraping, digging up, and replacing the sand. The first expense occurs in general monthly, and the second only once a year. At other places where I have secured corresponding data the figures range from 19 to 40 days' labor to scrape one acre, and average about the same as Lindley estimates.

Under some conditions sand-washing does not pay, and in

still others it is almost impossible. No apparatus has yet been devised which will wash the dirt out of the fine dune-sands used in Holland without washing a large part of the sand itself away, and in these works fresh sand, which is available in unlimited quantities and close to the works, is always used. At Breslau the dirty sand is sold for building purposes for one third of the price paid for new sand dredged from the river, delivered at the works, and no sand is ever washed. Budapest, Warsaw, and Rotterdam also use fresh river-sand without washing, except a very crude washing to remove clay at Budapest.

CHAPTER VI.

THEORY AND EFFICIENCY OF CONTINUOUS FILTRATION.

THE first filters for a public water-supply were built by James Simpson, engineer of the Chelsea Water Company at London in 1839. They were apparently intended to remove dirt from the water in imitation of natural processes, and without any very clear conception of either the exact extent of purification or the way in which it was to be accomplished. The removal of turbidity was the most obvious result, and a clear effluent was the single test of the efficiency of filtration, as it remains the legal criterion of the work of the London filters even to-day, notwithstanding the discovery and use of other and more delicate tests.

The invention and use of methods for determining the organic matters in water by Wanklyn and Frankland, about 1870, led to the discovery that the proportion of organic matters removed by filtration was disappointingly low, and as, at the time, and for many years afterward, an exaggerated importance was given to the mere quantities of organic matters in water, it was concluded that filtration had only a limited influence upon the healthfulness of the filtered water, and that practically as much care must be given to securing an unpolluted water as would be the case if it were delivered direct without filtration. This theory, although not confirmed by more recent investigation, undoubtedly has had a good influence upon the English works by causing the selection of raw waters free from excessive pollutions, and, in cases like the London supplies, drawn from the Thames and the Lea, in stimulating a most jealous care of the watersheds and the purification of sewage by the towns upon them.

It was only after the discovery of the bacteria in water and their relations to health that the hygenic significance of filtration commenced to be really understood. Investigations of the bacteria in the waters before and after filtration were carried out at Berlin by Plagge and Proskower, at London by Dr. Percy Frankland, and also at Zürich, Altona, and on a smaller scale at other places. These investigations showed that the bacteria were mainly removed by filtration, the numbers in the effluents rarely exceeding two or three per cent of those in the raw water. This gave a new aspect to the problem.

It was further observed, especially at Berlin and Zürich, that the numbers of bacteria in effluents were apparently quite independent of the numbers in the raw water, and the theory was formed that all of the bacteria were stopped by the filters, and that those found in the effluents were the result of contamination from the air and of growths in the underdrains. The logical conclusion from this theory was that filtered water was quite suitable for drinking regardless of the pollution of its source.

It was, however, found that the numbers of bacteria in the effluents were higher immediately after scraping than at other times, and it was concluded that before the formation of the sediment layer some bacteria were able to pass the sand, and it was therefore recommended that the first water filtered after scraping should be rejected.

Piefke at Berlin gave the subject careful study, and came to the conclusion that it was almost entirely the sediment layer which stopped the bacteria, and that the bacteria themselves in the sediment layer formed a slimy mass which completely intercepted those in the passing water. When this layer was removed by scraping, the action was stopped until a new crop of bacteria had accumulated. In support of this idea he stated that he had taken ordinary good filter-sand and killed the bacteria in it by heating it, and that on passing water through, no purification was effected—in fact, the effluent contained more bacteria than the raw water. After a little, bacteria established themselves in the sand, and then the usual purification was obtained. Piefke concluded that the action of the filter was a biological one; that simple straining was quite inadequate to produce the results obtained; that the action of the filter was mainly confined to the sediment layer, and that the depth of sand beyond the slight depth necessary for the support of this layer had no appreciable influence upon the results. The effect of this theory is still seen in the shallow sand layers used at Berlin and some other German works, although at London the tendency is rather toward thicker sand layers.

Piefke's deductions, however, are not entirely supported by his data as we understand them in the light of more recent investigation. The experiment with sterilized sand has been repeatedly tried at the Lawrence Experiment Station with results which quite agree with Piefke's, but it has also been found that the high numbers, often many times as high as in the raw water, do not represent bacteria which pass in the ordinary course of filtration, but instead enormous growths of bacteria throughout the sand supported by the cooked organic matter in it. It has been repeatedly found that ordinary sand quite incapable of supporting bacterial growths, after heating to a temperature capable of killing the bacteria will afterwards furnish the food for most extraordinary numbers. A filter of such sand may stop the bacteria of the passing water quite as effectually as any other filter, but if so, the fact cannot be determined without recourse to special methods, on account of the enormous numbers of bacteria in the sand, a small part of which are carried forward by the passing water, and completely mask the normal action of the filter.

The theory that all or practically all of the bacteria are intercepted by the sediment layer, and that those in the effluent are the result of growths in the sand or underdrains, received two hard blows in 1889 and 1891, when mild epidemics of typhoid fever followed unusually high numbers of bacteria in the effluents at Altona and at Stralau in Berlin, with good evidence in each case that the fever was directly due to the water. Both of these cases came during, and as the result of, severe winter weather with open filters and under conditions which are now recognized as extremely unfavorable for good filtration.

As a result of the first of these epidemics a series of experiments were made at Stralau by Fränkel and Piefke in 1890. Small filters were constructed, and water passed exactly as in the ordinary filters. Bacteria of special kinds not existing in the raw water or effluents were then applied, and the presence of a very small fraction of them in the effluents demonstrated beyond a doubt that they had passed through the filters under the ordinary conditions of filtration. These experiments were afterwards repeated by Piefke alone under somewhat different conditions with similar results. The numbers of bacteria passing, although large enough to establish the point that some do pass, were nevertheless in general but a small fraction of one per cent of the many thousands applied.

This method of testing the efficiency of filters had already been used quite independently by Prof. Sedgwick at the Lawrence Experiment Station in connection with the purification of sewage, and has since been extensively used there for experiments with water-filtration.

Kümmel also found at Altona that while in the regular samples for bacterial examination, all taken at the same time in the day, there was no apparent connection between the numbers of bacteria in the raw water and effluents, by taking samples at frequent intervals throughout the twenty-four hours, as has been done in a more recent series of experiments, and allowing for the time required for the water to pass the filters, a well-marked connection was found to exist between the numbers of bacteria in the raw water and in the effluents.

The subject has more recently been studied in much detail at the Lawrence Experiment Station, and it now appears that the bacteria in the effluent from a filter are from two sources: directly from the filtered water, and from the lower layers of the filter and underdrains. Thus we may say:

Bacteria in effluent = Bacteria from underdrains $+ \dfrac{a}{100} \times$ bacteria in raw water,

where a is the per cent of bacteria actually passing the filter.

Both of these terms depend upon a whole series of complex and but imperfectly understood conditions. In general the bacteria from the underdrains are low in cold winter weather, often almost *nil*, while at Lawrence with water temperatures of 70 to 75 degrees, and over, in July and August, the numbers from this source may reach 200 or 300, but for the other ten months of the year rarely exceed 50 under normal conditions. In summer especially it seems to be greater at low than at high rates of filtration

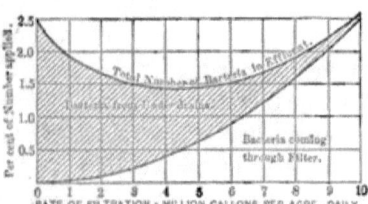

Fig. 14.—Showing Bacteria supposed to come through Filters and from the Underdrains.

(although a high rate for a short time only increases it), and so varies in the opposite way from the numbers actually passing the filters. This subject is by no means clearly understood; it is difficult, almost impossible, to separate the numbers of bacteria into the two parts—those which come directly through and may be dangerous, and those which have other origins and are harmless. The sketch, Fig. 14, is drawn to represent my idea of the way they may be divided. It has no statistical basis whatever. The light unshaded section shows the percentage number of bacteria which I conceive to be coming through a filter under given conditions at various rates of filtration, while the shaded section above represents the bacteria from other sources, and the upper line represents the sum of the two, or the total number of bacteria in the effluent. The relative importance of the two parts would probably vary widely with various conditions. With the conditions indicated by the sketch the number of bacteria in the effluent is almost constant: for a variation of only from 1.4 to 2.5 per cent of the number applied for the whole range is not a wide fluctuation for bacterial results, but the number in the lower and dangerous section is always rapidly increasing with increasing rate.

This theory of filtration accounts for many otherwise perplexing facts. The conclusion reached at Zürich and elsewhere that the efficiency of filtration is independent of rate may be explained in this way. This is especially probable at Zürich, where the number of bacteria in the raw water was only about 200, and an extremely large proportion relatively would have to pass to make a well-marked impression upon the total number in the effluent.

These underdrain bacteria are, so far as we know, entirely harmless; we are only interested in them to determine how far they are capable of decreasing the apparent efficiency of filtration below the actual efficiency, or the per cent of bacteria really removed by the filter.

This efficiency is dependent upon a large number of conditions many of which have already been discussed in connection with grain-size of filter sand, underdrains, rate of filtration, loss of head, etc., and a mere reference to them here will suffice. Perhaps the most important single condition is the rate, the numbers of bacteria passing increase rapidly with it. Next, fine sand and in moderately deep layers tends to give high efficiency. The influence of the loss of head, often mentioned, is not shown to be important by the Lawrence results, nor can I find satisfactory European results in support of it. Uniformity in the rate of filtration on all parts of the filtering area and a constant rate throughout the twenty-four hours are regarded as essential conditions for the best results. Severe winter weather has indirectly, by disturbing the regular action of open filters, an injurious influence, and has been the cause of most of the cases where filtered waters have been known to injure the health of those who have drunk them. This action is excluded in filters covered with masonry arches and soil, and such construction is apparently necessary for the best results in places subject to cold winters.

The efficiency of filtration under various conditions has been studied by a most elaborate series of experiments at Lawrence with small filters to which water has been applied containing a bacterium ($B.\ prodigiosus$) which does not occur naturally in this country and is not capable of growing in the filter, so that the results should represent only the bacteria coming through the filter and not include any additions from the underdrains. These results, which have been published in full in the reports of the Massachusetts State Board of Health, especially for the years 1892 and 1893, show that the number of bacteria passing increases rapidly with increasing rate, and slowly with decreasing sand thickness and increased size of sand-grain.

Assuming that the number of bacteria passing is expressed by the formula

$$\text{Per cent bacteria passing} = \frac{1}{2} \frac{(\text{rate})^2 \times \text{effective size of sand}}{\sqrt{\text{thickness of the sand in inches}}}$$

where the rate is expressed in million gallons per acre daily, and calculating by it the numbers of bacteria for the seventy-three months for which satisfactory data are available from 11 filters in 1892 and 1893, we find that

In 14 cases the numbers observed were 4 to 9 times as great as the calculated numbers;

In 6 cases they were 2 to 3 times as great;

In 35 cases they were between $\frac{1}{2}$ and 2 times the calculated numbers.

In 17 cases they were $\frac{1}{2}$ to $\frac{1}{8}$ of them.

In 11 cases they were less than $\frac{1}{8}$ the calculated numbers.

The agreement is only moderately good, and in fact no such formula could be expected to give more than very rough approximations, because it does not take into consideration the numerous other elements, such as uniformity and regularity of filtration, the influence of scraping, the character of the sediment in the raw water, etc., which are known to affect the results. Perhaps the most marked general difference is the tendency of new or freshly-filled filters to give higher, and of old and well-compacted filters to give lower, results than those indicated by the formula.

Comparing this formula with Piefke's results given in his "Neue Ermittelungen"* the formula gives in the first series (0.34 mm. sand, 0.50 m. thick, and rate 100 mm. per hour), 0.25 per cent passing, while the average number of *B. violacious* reported, excluding the first day of decreased efficiency after scraping, was 0.26 per cent. In the second series, with half as high a rate the numbers checked exactly the calculated 0.06 per cent.

In other experiments,† however, in 1893, when the calculated per cent was also 0.25, only 0.03, 0.04, and 0.07 per cent were observed in the effluents.

Comparing the results from the actual filters, (which numbers also include the bacteria from the underdrains and should therefore be somewhat higher) with the numbers calculated as passing through, I find that for the 46 days, Aug. 20 to Oct. 4, 1893, for which detailed results of the Stralau works are given by Piefke, the average calculated number passing is 0.20 per

---

* *Journal für Gas- und Wasserversorgung*, 1891, 108.
† *Zeitschrift für Hygiene*, 1894, 182.

cent, while twice as many were observed in the effluents; although three of the filters gave better effluents than the other eight, and the numbers from them approximated closely the calculated numbers. If we calculate the percentages of bacteria passing a number of filters, using the maximum rate of filtration allowed for the German filters where this is accurately determined, and for the English filters taking the maximum rate at one and one-half times the rate obtained by dividing the daily quantity by the area of filters actually in use, we obtain:

	Average Depth of Sand, Inches.	Effective Size of Sand-grain.	Maximum Rate of Filtration.	Per cent Bacteria passing $= \frac{1 \; r^{3d}}{2 \sqrt{sand}}$
Hamburg	32	0.31	1.60	0.07
Altona	28	0.34	2.57	0.21
Berlin, Stralau	20	0.34	2.57	0.25
" Müggel	20	0.34	2.57	0.25
" Tegel	20	0.37	2.57	0.27
London, Southwark & Vauxhall	36	0.34	2.81	0.22
" West Middlesex	39	0.37	2.81	0.23
" Chelsea	54	0.36	3.27	0.26
" Grand Junction	30	0.40	3.27	0.39
" Lambeth	36	0.36	3.75	0.42
Middlesborough	20	0.42	5.85	1.58
Zürich	26	0.35	7.50	1.90

The numbers actually observed are in every case higher than the calculated per cents passing, as indeed they should be on account of those coming from the underdrains, accidental contamination of the samples, etc.

It may be said that filtration as now practised in European works under ordinary conditions never allows over 1 or 2 per cent of the bacteria of the raw water to pass, and ordinarily not over one fourth to one half of one per cent, although exact data cannot be obtained owing to masking effect of the bacteria which come in from below and which bear no relation to those of the raw water. By increasing the size of filters, fineness and depth of sand (as at Hamburg), the efficiency can be materially increased above these figures. At the same time it must be borne in mind that the effectiveness of a filter may be greatly impaired by inadequate underdraining, by fluctuating rates of filtration where these are allowed, by freezing in winter in the case of open filters in cold climates, and by other irregularities, all of which can be prevented by careful attention to the respective points.

The action of a continuous filter throughout is mainly that of an exceedingly fine strainer, and like a strainer is mainly confined to the suspended or insoluble matters in the raw water. The turbidity, sediment, and bacteria of the raw water are largely or entirely removed, while hardness, organic matter, and color, so far as they are in solution, are removed to only a slight extent, if at all. Hardness can be removed by the addition of lime in carefully determined quantity before filtration (Clark's process), by means of which the excess of carbonic acid in the water is absorbed and the lime added, together with that previously in the water, is precipitated

Ordinary filtration will remove from one fourth to one third of the yellow-brown color of peaty water. A larger proportion can be removed by the addition of alum, which by decomposing forms an insoluble compound of alumina with the coloring matter, while the acid of the alum goes into the effluent either as free acid, or in combination with the lime or other base in the water, according to their respective quantities. Freshly precipitated alumina can be substituted for the alum at increased expense and trouble, and tends to remove the color without adding acid to the water. These will be discussed more in detail in connection with mechanical filters. Alum is but rarely used in slow sand filtration, the most important works where it is used being in Holland with peaty waters.

After all, the most conclusive test of the efficiency of filtration is the healthfulness of the people who drink the filtered water;

THEORY AND EFFICIENCY OF CONTINUOUS FILTRATION. 89

and the fact that many European cities take water-supplies from sources which would not be considered fit for use in the United States and, after filtering them, deliver them to populations having death-rates from water-carried diseases which are so low as to be the objects of our admiration, is the best proof of the efficiency of carefully conducted filtration.

It is only necessary to refer to London, drawing its water from the two small and polluted rivers, the Thames and the Lea; to Altona, drawing its water from the Elbe, polluted by the sewage of 6,000,000 people, 700,000 of them within ten miles above the intakes; to Berlin, using the waters of the Havel and the Spree; to Breslau, taking its water from the Oder charged with the sewage of mining districts in Silicia and Galicia, where cholera is so common; to Lawrence, with its greatly decreased death-rate since it has had filtered water, and to the hundred other places which protect themselves from the infectious matters in their raw waters by means of filtration. A few of these cases are described more in detail in Appendices V to IX, and many others in the literature mentioned in Appendix X.

An adequate presentation of even those data which have been already worked up and published would occupy too much space. I think every one who has carefully studied the recent history of water filtration in its relation to disease has been convinced that filtration carefully executed under suitable and normal conditions, even if not an absolute, is at least a very substantial protection against water-carried diseases, and the few apparent failures to remove objectionable qualities have been without exception due to abnormal conditions which are now understood and in future can be prevented.

BACTERIAL EXAMINATION OF WATERS.

Every large filter-plant should have arrangements for the systematic bacterial examination of the water before and after filtration, especially where the raw water is subject to serious pollution. Such examinations need not be excessively expensive, and they will not only show the efficiency of the plant as a whole, but may be made to show the relative efficiencies of the separate filters, the relative efficiencies at different parts of the periods of operation, the effect of cold weather, etc., and will then be a substantial aid to the superintendent in always securing good effluents at the minimum cost.

In addition a complete record of the bacteria in the water at different times may aid in determining definitely whether the water was connected with outbreaks of disease. Thus if an outbreak of disease of any kind were preceded at a certain interval by a great increase in the number of bacteria,—as has been the case, for example, with the typhoid epidemics at Altona and Berlin (see Appendices II and VII),—a presumption would arise that they might have been connected with each other, and each time it was repeated the presumption would be strengthened, while, on the other hand, outbreaks occurring while the bacteria remained constantly low would tend to discredit such a theory.

Bacterial investigations inaugurated after an epidemic is recognized, as has frequently been done, seldom lead to results of value, both because the local normal bacterial conditions are generally unknown at the commencement of the investigation, and because the most important time, the time of infection, is already long past before the first samples are taken. The fact that such sporadic activities have led to few definite results should throw no discredit upon continued observations, which have repeatedly proved of inestimable value.

Considerable misconception of the use of bacterial examinations exists. The simple bacterial count ordinarily used, and of which I am now speaking, does not and cannot show whether a water contains disease-germs or not. I object to the Chicago water, not so much because a glass of it contains a hundred thousand bacteria more or less, as because I am convinced, by a study of its source in connection with the city's death-rate, that it actually carries disease-germs which prove injurious to thousands of those who drink it. Now the fact being admitted that the water is injurious to health, variations in the numbers of bacteria in the water drawn from different intakes and at different times probably correspond roughly with varying proportions of fresh sewage, and indicate roughly the relative dangers from the use of the respective waters. If filters should be introduced, the numbers of bacteria in the effluents under various conditions would be an index of the respective efficiencies of filtration, and would serve to detect poor work, and would probably suggest the measures necessary for better results.

I would suggest the desirability of such investigations where mechanical filters are used, quite as much as in connection with slow filtration; and it would also be most desirable in the case of many water-supplies which are not filtered at all. Such continued observations have been made at Berlin since 1884; at London since 1886; at Boston and Lawrence since 1888; and recently at a large number of places, including Chicago, where observations by the city were commenced in 1894. They are now required by the German Government in the case of all filtered public water-supplies in Germany, without regard to the source of the raw water. The German standard requires that the effluent from each single filter, as well as the mixed effluent and raw water, shall be examined daily, making at some works 10 to 30 samples daily. This amount of work, however, can usually be done by a single man; and when a laboratory is once started, the cost of examining 20 samples a day will not be much greater than if only 20 a week are taken. In England and at some of the Continental works drawing their waters from but slightly polluted sources, much smaller numbers of samples are examined.

The question whether the examinations should be made under the direction of the water-works company or department, or by an independent body—as, for instance, by the Board of Health—

will depend upon local conditions. The former arrangement gives the superintendent of the filters the best chance to study their action, as he can himself control the collection of samples in connection with the operation of the filters, and arrange them to throw light upon the points he wishes to investigate; while examination by a separate authority affords perhaps greater protection against the possible carelessness or dishonesty of water-works officials. An arrangement being adopted in many cases in Germany is to have a bacterial laboratory at the works which is under the control of the superintendent, and in which the very numerous compulsory observations are made, while the Board of Health causes to be examined from time to time by its own representatives, who have no connection with the water-works, samples taken to check the water-works figures, as well as to show the character of the water delivered.

It seems quite desirable to have a man whose principal business is to make these examinations; as in case he also has numerous other duties, the examinations may be found to have been neglected at some time when they were most wanted. Such a man should have had thorough training in the principles of bacterial manipulation, but it is quite unnecessary that he should be an expert bacteriologist, especially if a competent bacteriologist is retained for consultation in cases of doubt or difficulty.

CHAPTER VII.

INTERMITTENT FILTRATION.

By intermittent filtration is understood that filtration in which the filtering material is systematically and adequately ventilated, and where the water during the course of filtration is brought in contact with air in the pores of the sand. In continuous filtration, which alone has been previously considered, the air is driven out of the sand as completely as possible before the commencement of filtration, and the sand is kept continuously covered with water until the sand becomes clogged and a draining, with an incidental aeration, is necessary to allow the filter to be scraped and again put in service.

In intermittent filtration, on the other hand, water is taken over the top of the drained sand and settles into it, coming in contact with the air in the pores of the sand, and passes freely through to the bottom when the water-level is kept well down. After a limited time the application of water is stopped, and the filter is allowed to again drain and become thoroughly aerated preparatory to receiving another dose of water.

This system of treating water was suggested by the unequalled purification of sewage effected by a similar treatment. It has been investigated at the Lawrence Experiment Station, and applied to the construction of a filter for the city of Lawrence, both of which are due to the indefatigable energy of Hiram F. Mills, C.E.

In its operation intermittent differs from continuous filtration in that the straining action is less perfect, because the filters yield no water while being aerated, and must therefore filter at a greater velocity when in use to yield the same quantity of water in a given time, and also on account of the mechanical disturbance which is almost invariably caused by the application of the water; but, on the other hand, the oxidizing powers of the filter, or the tendency to nitrify and destroy the organic matters, are stronger, and in addition, if the rate is not too high, the bacteria die more rapidly in the thoroughly aerated sand than is the case with ordinary filters.

It was found at Lawrence in connection with sewage filters that when nitrification was actively taking place the numbers of bacteria were much lower than under opposite conditions, and it was thought that nitrification in itself might cause the death of the bacteria. Later experiments, however, with pure cultures of bacteria of various kinds applied to intermittent filters with water to which ammonia and salts suitable for nitrification were added, showed that bacteria of all the species tried were able to pass the filter in the presence of nitrification, producing at least one thousand times as much nitrates as could result in any case of water-filtration, as freely as was the case when the ammonia was not added and there was but little nitrification. These results showed conclusively that nitrification in itself is not an important factor in bacterial removal, although nitrification and bacterial purification do to some extent go together; perhaps in part because the nitrification destroys the food of the bacteria and so starves them out, but probably much more because the conditions of aeration, temperature, etc., which favor nitrification also favor equally, and even in its absence, the death of the bacteria.

The rate at which water must pass through an intermittent filter is, on account of the intervals of rest, considerably greater than that required to give a corresponding total yield from a continuous filter, and its straining effect is reduced to an extent comparable to this increase in rate; and if other conditions did not come in, the bacterial efficiency of an intermittent filter would remain below that of a continuous one.

As a matter of fact the bacterial efficiency has usually been found to be less with intermittent filters at the Lawrence Experiment Station, when they have been run at rates such as are commonly used for continuous filters in Europe, say from one and one half to two million gallons and upwards per acre daily. With lower rates, and especially with rather fine materials, the bacterial efficiency is much greater; but it may be doubted whether it would ever be greater than that of a continuous filter with the same filtering material and the same total yield per acre. The number of bacteria coming from the underdrains is apparently generally less, and with very high summer temperatures much less, than in continuous filters, and this often gives an apparent bacterial superiority to the intermittent filters.

The effluents from intermittent often contain less slightly organic matter than those from continuous filters; but, on the other hand, hardly any water proposed for a public water-supply has organic matter enough to be of any sanitary significance whatever, apart from the living bodies which often accompany it; and if the latter are removed by straining or otherwise, we can safely disregard the organic matters. In addition, the water filtered will in a great majority of cases have enough air dissolved in itself to produce whatever oxidation there is time for in the few hours required for it to pass the filter, and it is only at very low rates of filtration that intermittent filters produce effluents of greater chemical purity than by the ordinary process. The yellow-brown coloring matter present in so many waters appears to be quite incapable of rapid nitrification; and where it is to some extent removed by filtration, the action is dependent upon other and but imperfectly understood causes which seem to act equally in continuous and intermittent filters.

The peculiarities of construction involved by this method of filtration will be best illustrated by a discussion of the Lawrence city filter designed by Hiram F. Mills, C.E., which is the only filter in existence upon this plan.*

---

* I am informed that several other filters upon the same principle have been more recently built.

THE LAWRENCE FILTER.

The filter consists of a single bed 2½ acres in area, the bottom of which is 7 feet below low water in the river, and filled with gravel and sand to an average depth of 4½ feet. The filter is all in a single bed instead of being divided into the three or four sections which would probably have been used for a continuous filter of this size. The water-tight bottom also was dispensed with, and the gravel was prevented from sinking into the silt by thin intermediate layers of graded materials. The saving in cost was considerable; but, on the other hand, a considerable quantity of ground-water comes up through the bottom and increases the hardness of the water from 1.5 to 2.6 parts of calcium carbonate in 100,000; and while the water when compared with many other waters is still extremely soft, the addition cannot be regarded as desirable. The ground-water also contains iron, which increases the color of the water above what it would otherwise be.

The underdrains have a frictional resistance ten times as great as would be desirable for a continuous filter, the idea being to check extreme rates of filtration in case of unequal flooding, and also to limit the quantity of water which could be gotten through the filter to that corresponding to a moderate rate of filtration.

The sand, instead of being all of the same-sized grain, is of two grades, with effective sizes respectively 0.25 and 0.30 mm., the coarser sand being placed farthest away from the underdrains, where its greater distance is intended to balance its reduced frictional resistance and make all parts filter at an equal rate.

The surface instead of being level is waved, that is, there are ridges thirty feet apart, sloping evenly to the valleys one foot deep half way between them, to allow water to be brought on rapidly without disturbing the sand surface. For the same reason, as well as to secure equality of distribution, a system of concrete carriers for the raw water goes to all parts of the filter, reducing the effective filtering area by 4 or 5 per cent. The filter is scraped as necessary in sections, the work being performed when the filter is having its daily rest and aeration. Owing to the difference in frictional resistance before and after scraping, and to the fact that it is impossible to scrape the entire area in one day, considerable variations in the rate of filtration in different parts of the filter must occur. The heavy frictional resistance of the underdrains when more than the proper quantity of water passes them tends to correct this tendency especially for the more remote parts of the filter, but perhaps at the expense of those near to the main drain.

The filter is not covered as the suggestions in Chapter II would require, but this is hardly on account of its being an intermittent filter.

The annual report of the Massachusetts State Board of Health for 1893 states that during the first half of December, 1893, the surface remained covered, that is, it was used continuously, and after December 16th it was so used when the temperature was below 24°, and was drained only when the temperature was 24° or above. The days on which the filter was drained during the remainder of December are not given, but during January and February, 1894, the filter remained covered 29 days and was drained 30 days. Bacterial samples were taken on 44 of these days, 22 days when it was drained and 22 when it was not. The average number of bacteria on the days when it was not drained was 137 and on those days when it was drained 252 per cubic centimeter.

From February 24th to March 12th the number of bacteria were unusually high, averaging 492 per cubic centimeter, or 5.28 per cent of the 9308 applied. During this period the filter was used intermittently; there was ice upon it, and parts of the surface were scraped under the ice, and high rates of filtration undoubtedly resulted on the scraped areas. After March 12th the ice had disappeared and very much better results were obtained.

While there may be some question as to the direct cause of this decreased efficiency with continued cold weather and ice, the results certainly are not such as to show the advisability of building open filters in the Lawrence climate.

The cost of building the filter in comparison with European filters was extraordinarily low—only $67,000, or $27,000 per acre of filter surface. To have constructed open continuous filters of the same area with water-tight bottoms, divided into sections with separate drains and regulating apparatus, with the necessary piping, would have cost at least half as much more, and with the masonry cover which I regard as most desirable in the Lawrence climate the cost would have been two or three times the expenditure actually required.

It was no easy matter to secure the consent of the city government to the expenditure of even the sum used; there was much skepticism as to the process of filtration in general, and it was said that mechanical filters could be put in for about the same cost. Insisting upon the more complete and expensive form might have resulted either in an indefinite postponement of action, or in the adoption of an inferior and entirely inadequate process. Still I feel strongly that in the end the greater expense would have proved an excellent investment in securing softer water and in the greater facility and security of operating the filter in winter.

In regard to the effect of the Lawrence filter upon the health of the city, I can best quote from Mr. Mills' paper in the Report of the Massachusetts State Board of Health for 1893, and also published in the Journal of the New England Water-works Association. Mr. Mills says: " In the following diagram [Fig. 15] the average number of deaths from typhoid fever at Lawrence for each month from October to May, in the preceding five years, are given by the heavy dotted line; and the number during the past eight months are given by the heavy full line.

"The total number for eight months in past years has been forty-three, and in the present year seventeen, making a saving of twenty-six. Of the seventeen who died nine were operatives

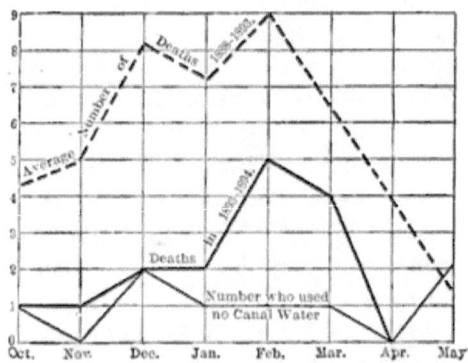

FIG. 15.—TYPHOID FEVER IN LAWRENCE.

in the mills, each of whom was known to have drunk unfiltered canal water, which is used in the factories at the sinks for washing.

"The finer full line shows the number of those who died month after month who are not known to have used the poisoned canal water. The whole number in the eight months is eight.

"It is evident from the previous diagram [not reproduced] that the numbers above the fine full line, here, follow after those at Lowell in the usual time, and were undoubtedly caused by the sickness at Lowell; but we have satisfactory reason to conclude that the disease was not propagated through the filter but that the germs were conveyed directly into the canals and to those who drank of the unfiltered canal water. Among the operatives of one of the large corporations not using the canal water there was not a case of typhoid fever during this period. Warnings have been placed in the mills where canal water is used to prevent the operatives from drinking it.

"We find, then, that the mortality from typhoid fever has, during the use of the filter, been reduced to 40 per cent of the former mortality, and that the cases forming nearly one half of this 40 per cent were undoubtedly due to the continued use of unfiltered river water drawn from the canals."

The results for the remainder of 1894 have been equally favorable, and show conclusively the value of filtration.

CHEMNITZ WATER-WORKS.

The only other place which I have found where anything approaching intermittent filtration of water is systematically employed is Chemnitz, Germany. The method there used bears the same relation to intermittent filtration as does broad irrigation of sewage to the corresponding method of sewage treatment; that is, the principles involved are mainly the same, but a much larger filtering area is used, and the processes take place at a lower rate and under less close control.

The water-works were built about twenty years ago by placing thirty-nine wells along the Zwönitz River, connected by siphon pipes, with a pumping-station which forced the water to an elevated reservoir near the city (Fig. 16). The wells are built of masonry, 5 or 6 feet in diameter and 10 or 12 feet deep, and are on the rather low bank of the river. The material, with the exception of the surface soil, and loam about 3 feet deep, is a somewhat mixed gravel with an effective size of probably from 0.25 to 0.50 mm., so that water is able to pass through it freely. The wells are, on an average, about 120 feet apart, and the line is seven eighths of a mile long.

INTERMITTENT FILTRATION.

It was found that in dry times the ground-water level in the entire neighborhood was lowered some feet below the level of the river without either furnishing water enough or stopping the flow of the river below. The channel of the river was so silted that, notwithstanding the porous material, the water could not penetrate it to go toward the wells.

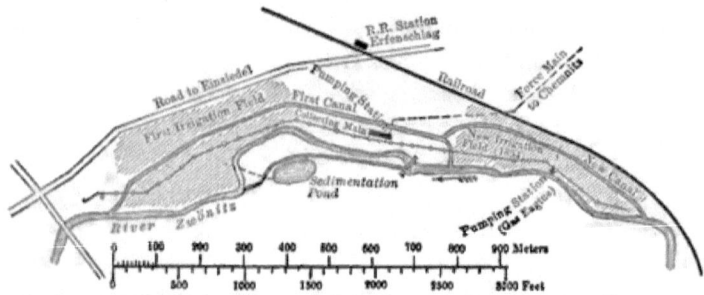

FIG. 16.—PLAN OF AREA USED FOR INTERMITTENT FILTRATION AT CHEMNITZ.

A dam was now built across the river near the pumping-station, and a canal was dug from above the dam, crossing the line of wells and running parallel to it on the back side for about half a mile. Later a similar canal was dug back of the remaining upper wells. Owing to the difference in level in the river above and below, the canals can be emptied and filled at pleasure. They are built with carefully prepared sand bottoms, and the sand sides are protected by an open paving, to allow the percolation of as much water as possible, and the sand is cleaned by scraping, as is usual with ordinary sand filters, once a year or oftener.

The yield from the wells was much increased by these canals, but the water of the river is polluted to an extent which would ordinarily quite prevent even the thought of its being used for water-supply, and it was found that the water going into the ground from the canals, and passing through the always saturated gravel to the wells, without coming in contact with air at any point, after a time contained iron and had an objectionable odor.

To avoid this disagreeable result the meadow below the pumping-station was laid out as an irrigation field (Fig. 16). The water from above the dam was taken by a canal on the opposite side of the river through a sedimentation pond (which, however, is not now believed to be necessary and is not always used), and then under the river by a siphon to a slightly elevated point on the meadow, from which it is distributed by a system of open ditches, exactly as in sewage irrigation. The area irrigated is not exactly defined and varies somewhat from time to time; the rate of filtration may be roughly estimated at from 100,000 to 150,000 gallons per acre daily, although limited portions may occasionally get five times these quantities for a single day. The water passes through the three feet of soil and loam, and afterward through an average of six feet of drained coarse sand or gravel in which it meets air, and afterward filters laterally through the saturated gravel to the wells. The water so obtained is invariably of good quality in every way, colorless, free from odor and from bacteria. The surface of the irrigated land is covered with grass and has fruit-trees (mostly apple) at intervals over its entire area.

This first system of irrigation is entirely by gravity. On account of natural limits to the land it could not be conveniently extended at this point, and to secure more area, the higher land above the pumping-station was being made into an irrigation field in 1894. This is too high to be flooded by gravity, and will be used only for short periods in extremely dry weather. The water is elevated the few feet necessary by a gas-engine on the river-bank. In times of wet weather enough water is obtained from the wells without irrigation, and the land is only irrigated when the ground-water level is too low.

During December, January, and February irrigation is usually impossible on account of temperature, and the canals are then used, keeping them filled with water so that freezing to the bottom is impossible; but trouble with bad odors in the filtered water drawn from the wells is experienced at these times.

The drainage area of the Zwönitz River is only about 44 square miles, and upon it are a large number of villages and factories, so that the water is excessively polluted. The water in the wells, however, whether coming from natural sources, or from irrigation, or from the canals, has never had as many as 100 bacteria per cubic centimeter, and is regarded as entirely wholesome.

In extremely dry weather the river, even when it is all used for irrigation so that hardly any flows away below, cannot be made to supply the necessary daily quantity of 2,650,000 gallons, and to supply the deficiency at such times, as well as to avoid the use of the canals in winter, a storage reservoir holding 95,000,000 gallons has recently been built on a feeder of the river. This water, which is from an uninhabited drainage area, is filtered through ordinary continuous filters and flows to the city by gravity. Owing to the small area of the watershed it is incapable of supplying more than a fraction of the water for the city, and will be used to supplement the older works.

This Chemnitz plant is of especial interest as showing the successful utilization of a river-water so grossly polluted as to be incapable of treatment by the ordinary methods. Results obtained at the Lawrence Experiment Station have shown that sewage is incapable of being purified by continuous filtration, the action of air being essential for a satisfactory result. With ordinary waters only moderately polluted this is not so; for they carry enough dissolved air to effect their own purification. In Chemnitz, however, as shown by the results with the canals, the pollution is so great that continuous filtration is inadequate to purify the water, and the intermittent filtration adopted is the only method likely to yield satisfactory results in such cases.

Intermittent filtration is now being adopted for purifying brooks draining certain villages and discharging into the ponds or reservoirs from which Boston draws its water-supply. The water of Pegan Brook below Natick has been so filtered since 1893 with most satisfactory results, and affords almost absolute protection to Boston from any infection which might otherwise enter the water from that town. A similar treatment is soon to be given to a brook draining the city of Marlborough. The sewage from these places is not discharged into the brooks, but is otherwise provided for, but nevertheless they receive many polluting matters from the houses and streets upon their banks.

The filtration used resembles in a measure that at Chemnitz, and I am informed by the engineer, Mr. Desmond FitzGerald, that it was adopted on account of its convenience for this particular problem, and not because he attaches any special virtue to the intermittent feature.

APPLICATION OF INTERMITTENT FILTRATION.

In regard to the use of waters as grossly polluted as the Zwönitz, the tendency is strongly to avoid their use, no matter how complete the process of purification may be; but in case it should be deemed necessary to use so impure a water for a public supply, intermittent filtration is the only process known which would adequately purify it. And it should be used at comparatively low rates of filtration. I believe that an attempt to filter the Zwönitz at the rate used for the Merrimac water at Lawrence, which is by comparison but slightly polluted, would result disastrously.

The operation in winter must also be considered. Intermittent filtration of sewage on open fields in Massachusetts winters is only possible because of the comparatively high temperature of the sewage (usually 40° to 50°), and would be a dismal failure with sewage at the freezing-point, the temperature to be expected in river-waters in winter.

It is impossible to draw a sharp line between those waters which are so badly polluted as to require intermittent filtration for their treatment and those which are susceptible to the ordinary continuous filtration. Examples of river-waters polluted probably beyond the limits reached in any American waters used for drinking purposes and successfully filtered with continuous filters are furnished by Altona, Breslau, and London.

Intermittent filtration may be considered in those cases where it is proposed to use a water polluted entirely beyond the ordinary limits, and for waters containing large quantities of decomposable organic matters and microscopical organisms; but in those cases where a certain and expeditious removal of mud is desired, and where waters are only moderately polluted by sewage, but still in their raw state are unhealthy, it is not apparent that intermittent filtration has any advantages commensurate with the disadvantages of increased rate to produce the same total yield and of the increased difficulty of operation, particularly in winter; and in such cases continuous filtration is to be preferred.

CHAPTER VIII.

OTHER METHODS OF FILTRATION.

MECHANICAL FILTERS WITHOUT COAGULENTS.

THE mechanical filters so largely used in the United States to clarify water for manufacturing purposes consist, in general, of iron cylinders filled with sand through which water is forced at rates of from one to three hundred million gallons per acre daily, or we may say, in a general way, one hundred times as fast as in the European filters for public water-supplies. These filters, of which there are many patented kinds upon the market, differ from each other mainly in the method with which the clogging of the sand is removed. Instead of scraping, as in the case of the slow sand-filters at intervals of some weeks, the whole body of the sand is washed in the filter itself at short intervals, depending upon the rapidity of clogging.

I wish to emphasize specially the fact that the subject of the various patents and the differences between the different makes of filters and between the whole class of mechanical filters and slow sand-filters lie in the methods of cleaning the sand and regulating the rate, pressure, etc. So far as purification is concerned, the principles of these various filters are identical with each other and with those governing the action of slow filters. The failure to grasp this fundamental point has led to not a little misunderstanding in regard to them.

Mechanical filters were originated in paper-mills to remove from the vast volumes of water required those comparatively large particles which would otherwise affect the appearance or texture of the paper. Exactly how large a particle must be to injure the appearance of paper I do not know; but I can hardly think that anything less than one thousandth of an inch in diameter would be objectionable, while the particles which cause turbidity in drinking-water as well as the germs of disease are often less than one tenth of this length. The cost of filtration which removes particles one ten-thousandth of an inch long is unquestionably greater than that which only removes particles one thousandth of an inch long; and if the latter suffices for paper-making, a manufacturer will have reason to be satisfied with it, and will not insist upon the more thorough and expensive treatment. The argument does not apply to a city water-supply, where the requirements are radically different.

From what has been said in Chapter IV it would follow that filtration at any such rates as are invariably followed in mechanical filters, and are necessary with them, on account of the high relative cost of the effective filtering area, would be quite incapable of removing the bacteria. And this is the case so long as the use of alum or other coagulents is not considered. I do not know of any evidence whatever that mechanical filters without the use of alum effect the removal of more than an unimportant fraction of the bacteria. My own experience with filters of this class has never shown the removal of more than from ten to fifty per cent of bacteria, and I regard this as a fair estimate of their bacterial efficiency in general, although it is undoubtedly true that samples taken just before the cleaning of the filter may show much better results.

Moreover, I have received direct evidence that polluted water filtered in this way is not rendered free from infection, but is capable of causing disease apparently as freely as the same water would do without filtration.

A village in Northern Vermont, has two water-supplies. About half the population is supplied by an aqueduct company from a pond which is comparatively free from pollution, and in addition the water is filtered through three filters of fine sand, at a rate of six million gallons per acre daily, or less. The filters are covered as a protection against frost. The other or public supply is drawn from a small river flowing through the villiage and from a point within the village limits. A town with 2000 inhabitants, is nine miles above, and another village with a considerable although smaller population, is only three miles away; and within the limits of the village itself there are a number of houses standing directly upon the banks of the river, and which undoubtedly drain into it, although there are no public sewers discharging into the river above the water intake. The obvious sources of pollution of the town supply caused anxiety in regard to its quality, and to improve the supply a mechanical filter of one of the leading makes was purchased, and was put in operation Sept. 10, 1892. No alum was used.

During the following winter and spring there was a severe epidemic of typhoid fever. With a population of about 6000 there were 150 cases and 30 deaths. The Health Officer found upon examination that of these, 135 cases were among those who were regularly supplied with the town water, and that of the 15 cases among the half of the population supplied by the aqueduct company a considerable number had access, at their places of business or elsewhere, to the town supply.

The epidemic was caused beyond a reasonable doubt by the use of the polluted river water-supply, and the fact that it was filtered by a mechanical filter, without alum, at a high rate, was no protection to the town.

It is, after all, not a great exaggeration to say that in filtering public water-supplies the use of a mechanical filter without coagulent only "takes the logs out," for it certainly fails to remove those substances which are most objectionable in drinking-water, and which are capable of being removed by better forms of filtration.

Two reasons are apparent which account to some extent for the use of this form of filtration for public water-supplies. First, the removal of the visible floating or suspended particles which are the most obvious foreign matters in the water, and the removal of which is too often taken by ignorant persons for an index of the removal of other matters which cannot be so readily observed; and, second, the sight of large quantities of dirt found in the filter after a run, and which were obviously removed from the water, implying that the latter became purified by losing so much filth. A little thought will show that this is exactly like arguing that because a pail of water was taken out of the pond therefore the pond is now dry.

The whole subject of mechanical filtration without coagulents may be summed up by saying that any water not suitable for drinking raw will still be almost equally objectionable when filtered in this way.

THE USE OF ALUM.

The addition of a small quantity of alum or sulphate of alumina to water before filtration is often used in connection with mechanical filtration, and introduces an entirely new factor.

The alum is decomposed into its component parts, sulphuric acid and alumina, the former of which combines with the lime or other base present in the water, or, in case enough of this is lacking, it remains partly as free acid and partly as undecomposed alum, while the alumina forms a gelatinous precipitate which draws together and surrounds the suspended matters present in the water, including the bacteria, and allows them to be much more easily removed by filtration. In addition, the alumina in some way which is not understood has a chemical attraction for dissolved organic matters, and the chemical purification and removal of color with the use of alum may be more complete at very high rates than would be possible at any rate, however low, with simple filtration.

The use of alum is neither new nor peculiar to mechanical filters. As early as 1831 D'Arcet published in the "Annales d'hygiène publique" * an account of the purification of Nile water in Egypt by adding alum to the water, and afterward filtering it through small household filters. More recently alum has been repeatedly used in connection with slow filters, particularly at Leeuwarden, Groningen, and Schiedam in Holland, where the river waters used for public supplies are colored by peaty matter which cannot be removed by simple filtration.

Its use has, however, generally been abandoned, or at least restricted to times when the raw water is unusually bad, either on account of its cost, or because it was unnecessary, carefully conducted filtration without it yielding as good an effluent as was necessary, or because injurious effects followed its use. Antwerp furnished a striking case of the latter in 1893. During extremely dry weather the much polluted river from which the water is drawn became extremely foul, and with the small filtering area available (since then increased) alum was resorted to to secure a better effluent, especially as there was some fear of cholera. The water left the filter quite clear, and free from color, undecomposed alum, or iron, and with a neutral reaction; but it apparently contained at least a trace of free acid, for after passing through some miles of pipe on its way to the town, it contained a considerable quantity of iron which it had taken from the pipe, and in addition it was objectionable in its appearance and quite unsuitable for laundry purposes, and caused much complaint and annoyance to the water company.

But while alum has been mainly discarded in European slow filters, it has been adopted in the American mechanical filters, and the favorable results which are cited in connection with them are usually obtained by its use.

The commonest objections to the use of alum are, that it is liable to leave the water acid, giving rise to troubles like those mentioned above; but this depends entirely upon the alkalinity

---

* Translation in German in Dingler's Polytechnical Journal, 1832, 386.

of the raw water and the proportion of alum used, and trouble of this sort could probably be satisfactorily corrected by adding a small quantity of soda to the water; and, second, that the alum remaining in the effluent is injurious to health. It is frequently claimed that the decomposition of the alum is complete, and that none of the alumina goes into the effluent; but the most careful investigation seems to show that with the rates of filtration necessarily employed with this system a certain although extremely small quantity of alum or alumina invariably remains in the effluent. Although alum in large quantity is undoubtedly injurious to health, it is neither a violent nor a cumulative poison; and the proposition that one part of alumina in a million parts of water is injurious to health must be regarded as conjecture rather than as a matter of proof, or even of probability.

In regard to the bacterial efficiency of mechanical filters with alum, we shall be fairly safe in concluding from analogy that low (comparative) rates are safer than higher rates, that fine sand and in a thick layer will be safer than the reverse conditions. Experiments have also shown that the amount of alum used affects the result, and that the same results cannot be obtained with a small dose as with a larger quantity. We can also assume that the bacterial efficiency will increase as the filter becomes clogged, and that much poorer results will be obtained immediately after washing than at other times. The statements of tests frequently given out which do not take these important factors into consideration can be safely neglected, as throwing no light upon the general problem. It is not a difficult matter for an intelligent person by skilful manipulation to produce a limited quantity of good effluent from almost any filter within reason, and the fact that a good effluent is so produced is of no interest whatever. Such results only have value when they are known to represent the total or some definite part of a filter's work when running under known conditions at least approximating those of actual practice.

A serious objection to the use of rapid filters with alum for polluted waters is the possibility of poor work resulting from the failure at some time to apply the alum. The success of the process from a sanitary standpoint depends absolutely upon the alum being applied to all the water and in the proper proportion. The filters will probably run day and night and every day in the year. The failure to apply alum, say for a few minutes in the middle of the night, means that for a time unpurified water will be delivered. The first water filtered after cleaning also is not free from bacteria, and must be allowed to go to waste. Sometime, when there is a scarcity of water or through carelessness, the first filtrate may be turned into the town supply with disastrous results.

How much damage may be caused by such accidents it is impossible to say; but after giving all possible credit to automatic machinery and faithful attendance, it must be said that they are much more likely to occur with the rapid mechanical filters than with slow sand-filters. With the latter, the various conduits can be so arranged that it is practically impossible for a negligent employé to turn raw water into the pure-water conduit; and even if he deliberately attempted to do so, it would be necessary for him to dig up the sand or make some corresponding violent change, which could not fail to be detected. The water which has actually penetrated the sand at the rates at which it will be possible for it to pass through the outlet openings will be sure under all conditions to be at least reasonably well purified even if the watchman stupidly opens the gates wide when a filter is first put in service, and such an error would be instantly and unmistakably apparent to any one familiar with the normal working of the filters. With a mechanical filter, on the other hand, it is easy to imagine that the alum solution might be exhausted without attracting attention, or that a lazy attendant might fail to replace it promptly.

There are two things which should be clearly shown in regard to a mechanical filter before allowing its adoption for a polluted city supply: first, that it is capable of removing the bacteria under normal conditions of operation; and, second, that the mechanical arrangements are so perfect that the result will not be dependent upon the attendant to such an extent that mere negligence, perhaps during the night when detection is improbable, may subject the city to all the bad results of polluted water. We can wait until these questions are settled in the affirmative before considering the question as to the efficiency and cost of mechanical as compared to slow filtration without chemicals.

THE USE OF PRECIPITATED ALUMINA.

I do not know that this process has ever gone beyond the experimental stage, but its plausibility, and the fact that some little information has been collected in regard to it, require its mention. Instead of adding the alum solution to the water to decompose in direct contact with the matters it is to affect, the alumina of the solution is previously precipitated with soda, washed free or nearly free from the resulting sulphate of soda, and the flocculent precipitate then added to the water. By this procedure both the acidification of the effluent and the possibility of the passage of undecomposed alum are entirely avoided; but, on the other hand, the power of the precipitated alumina is very much less than that of a corresponding quantity of alum. Experiments made by this process have not shown good bacterial efficiency at the rates of filtration followed in mechanical filters.

THE USE OF OTHER SALTS.

Ferric salts can be used in place of alum, but are open to the same objections and I do not know of their having been extensively employed. Cuprous chloride has been suggested by

Kröhnke,* who obtained very satisfactory bacterial results in some experiments made at Hamburg; but, if only for sentimental reasons, the process is not likely to be widely adopted.

THE USE OF METALLIC IRON.

The use of metallic iron for water purification in connection with a moderately slow filtration through filters of the usual form is known as Anderson's process (patented), and has been used at Antwerp and elsewhere on a large scale, and has been experimentally examined at a number of other places.

The process consists in agitating the water in contact with metallic iron, a portion of which is taken into solution as ferrous carbonate. Upon subsequent aeration this is supposed to become oxidized and precipitate out as ferric hydrate, with all the good and none of the bad effects of alum. The precipitate is partially removed by sedimentation, while filtration completes the process. The process is theoretically admirable, and in an experimental way upon a very small scale often gives most satisfactory results, muddy waters very difficult of filtration and colored peaty waters yielding promptly clean and colorless effluents. I have often personally made these experiments, and can vouch for the results, although failure repeatedly occurs under conditions apparently identical with those which at other times succeed brilliantly.

In applying the process on a larger scale, however, with peaty waters at least, it seems impossible to get enough iron to go into solution in the time which can be allowed, and the small quantity which is taken up either remains in solution or else slowly and incompletely precipitates out, without the good effects which follow the sudden and complete precipitation of a larger quantity, and in this case the color is seldom reduced, and may even be increased above the color of the raw water by the iron remaining in solution.

Journal für Gas- und Wasserversorgung, 1883, *p.* 513.

The ingenuity of those who have studied the process has not yet found any adequate means of avoiding these important practical objections; and even at Antwerp a great extension of the filtering area, as well as the use of alum at times of unusual pollution, is good evidence that simple filtration, in distinction from the effect of the iron, is relied upon much more than formerly.

At Dordrecht also, where the process has been long in use, the rate of filtration does not exceed the ordinary limits; nor is the result, so far as I could ascertain, in any way superior to that obtained a few miles away, at Rotterdam, by ordinary filtration, with substantially the same raw water.

The results obtained at Boulogne-sur-Seine, near Paris, have been closely watched by the public chemist and bacteriologist of Paris, and have been very favorable, and a new plant with a capacity of 7,900,000 gallons daily is just being built at Choisy-le-Roi, to supply some of the suburbs of Paris, but even in these cases only moderate rates of filtration are employed which would yield excellent effluents without the iron.

HOUSEHOLD FILTERS.

The subject of household filters is a somewhat broad one, as the variety in these filters is even greater than in the larger filters, and the range in the results to be expected from them is at least as great. I shall only attempt to indicate here some of the leading points in regard to them.

Household filters may be used to remove mud or iron rust from the tap water, or to remove the bacteria in case the latter is sewage-polluted, or to do both at once. Perhaps oftener they are used simply because it is believed to be the proper thing, and without any clear conception either of the desired result or the way in which it can be accomplished. I shall consider them only in their relations to the removal of bacteria, as I credit the people who employ them with being sufficiently good judges of their efficiency in removing visible sediment.

In the first place, as a general rule, which has very few if any exceptions, we may say that all small filters which allow a good stream of water to pass do not remove the bacteria. The reason for this is simply that a material open enough to allow water to pass through it rapidly is not fine enough to stop such small bodies as the bacteria. The filters which are so often sold as "germ-proof," consisting of sand, animal charcoal, wire-cloth, filter-paper, etc., do not afford protection against any unhealthy qualities which there may be in the raw water. Animal charcoal removes color without retaining the far more objectionable bacteria.

The other household filters have filtering materials of much finer grain, unglazed porcelain and natural sandstone being the most prominent materials, while infusorial earth is also used. The smaller sizes of these filters allow water to pass only drop by drop, and when a fair stream passes them the filters have considerable filtering area (as a series of filter-tubes connected together). On account of their slow action, filters of this class are, as a rule, provided with storage reservoirs so that filtered water to the capacity of the reservoir can be drawn rapidly (provided the calls do not come too often). Some of these filters are nearly germ-proof, and are comparable in their efficiency to large sand-filters. There is no sharp line between the filters which stop and which do not stop the bacteria; but in general the rule that a filter which works rapidly in proportion to its size does not do so, and *vice versa*, will be found correct.

In thinking of the efficiency of household filters we must distinguish between the filter carefully prepared for an award at an exhibition and the filter of the same kind doing its average daily work in the kitchen. If we could be sure in the latter case that an unbroken layer of fine sandstone or porcelain was always between ourselves and the raw tap-water we could feel comparatively safe. The manufacturers of the filters claim that leaky joints, cracked tubes, etc., are impossible; but I would urge upon the people using water filtered in this way that they personally assure themselves that this is actually the case with their own filters, for in case any such accident should happen the consequences might be most unpleasant. The increased yield of a filter due to a leaky joint is sure not to decrease it in favor with the cook, who is probably quite out of patience with it because it works so slowly, that is, in case it is good for anything.

The operation of household filters is necessarily, with rare exceptions, left to the kitchen-girl and luck. Scientific supervision is practically impossible. With a large filter, on the other hand, concentrating all the filters for the city at a single point, a competent man can be employed to run them in the best-known way; and if desired, and as is actually done in very many places, an entirely independent bacteriologist can be employed to determine the efficiency of filtration. With the methods of examination now available, and a little care in selecting the times and places of collecting the samples, it is quite impossible for a filter-superintendent to deliver a poor effluent very often or for any considerable length of time without being caught. The safety of properly-conducted central filtration is thus infinitely greater than that from even the best household filters. Further, it may be doubted whether an infected water can be sent into every house in the city to be used for washing and all the purposes to which water is put except drinking, without causing disease, although less than it would if it were also used for drinking.

The use of household filters must be regarded as a somewhat desperate method of avoiding some of the bad consequences of a polluted water-supply, and they are adopted for the most part by citizens who in some measure realize the dangers from bad water, but who cannot persuade their fellow-citizens to a more thorough and adequate solution of the problem. Such citizens, by the use of the best filters, and by carefully watching their action, or by having their drinking-water boiled, can avoid the principal dangers from bad water, but their vigilance does not protect their more careless neighbors.

CHAPTER IX.

COST AND ADVANTAGES OF FILTRATION.

THE cost of filtration of water by the methods now used in Europe if adopted in the United States would depend so largely upon local conditions that any accurate general estimate is quite impossible. Nevertheless a little consideration of the subject, although not leading to exact results, may be helpful as furnishing a rough idea of the probable cost before estimates for local conditions are made.

In England not much is to be learned in regard to the cost of filtration, because as a rule no separate accounts are kept of the filters. In Germany such accounts are generally kept, and many results are available. The average cost of building open filters with all the latest improvements, with the necessary reservoirs for filtered water, sand-washing apparatus, etc., but not including sedimentation-basins or the cost of land, is estimated by Lindley, an engineer of very wide experience, at from $42,000 to $48,000 per acre of effective filtering area, while for corresponding covered filters he estimates the cost at $63,000 to $73,000 per acre.

The total cost of filtering water, that is, the operating expenses, with 4 per cent on the capital invested for interest and 2 per cent for sinking funds, and 10 per cent depreciation on machinery, is at Zürich with clear lake-water only 0.65 cent per thousand gallons, but in general is from 0.7 to 1.3 cents, of which from one third to one half consists of the operating expenses, and the remainder is for the various charges for capital and depreciation.

The actual building costs of some German and other works have been as follows:

Place.	Date of Construction.	Cost per Acre. Covered.	Cost per Acre. Open.
Stralau	1874	$62,000	....
Tegel, first part	1884	66,000	....
" second "	1887	70,000	....
Hamburg	1892–3	....	67,000*
Königsberg, first part	1883	....	20,000
" second "	1889	39,000	....
Magdeburg	1876	83,000	....
Warsaw	1885	78,000	....
Zürich	1885	86,000	....
Lawrence	1892–3	....	27,000
Nantucket	1892	....	45,500

Let us consider the case of a city in the United States with 100,000 inhabitants now drawing its water from a polluted lake or river. The quantity of water required is perhaps 80 or 100 gallons per inhabitant daily. Let us assume an average consumption of 8,000,000 gallons daily and an ordinary maximum of 10,000,000 gallons. Half this quantity would be considered an ample allowance in England, and a still smaller quantity would be required by Continental cities of this size. Why do American cities require so much more water than European cities? If there was any reason to believe that the enormous volumes of water used in America contributed to the health, cleanliness, or comfort of those who use or waste them, we should not wish to reduce them, but many European cities are clean beyond the dreams of an American alderman, are bountifully supplied with public fountains, and water is supplied abundantly and at low rates to all inhabitants for domestic and manufacturing purposes. And still the per-capita consumption is less than a third of the American requirement. But this is not the place for a discussion of the quantity of water required; the existing condition must be either met or changed.

---

* Includes cost of large sedimentation-basins, pumps, and conduits, etc.

To supply a maximum of 10,000,000 gallons daily, five filters each with an area of one acre will be ample. Any four of them can easily furnish this quantity while the fifth is out of use for cleaning or other cause. If the city is north of the line of normal January temperature of 32° (page 17), vaulted filters will be required. Such filters have formerly cost on an average $68,000 per acre including everything in Germany, but the vaultings of these filters were often excessively thick, and the rather small size of the single beds increased the proportion of walls, regulators, piping, etc. Lighter vaulting has been used on some new filters for which I have no statements of cost.

Some estimates recently made by the author in connection with engineers examining the Boston Metropolitan Water-supply indicate that filters fully up to the German standards, but with beds of a full acre each, and with vaulting substantially like that successfully used on the Newton covered reservoir, can be built at present American prices for somewhat less than the cost given above, notwithstanding the higher price paid for American labor.

Including the connection with the (existing) pumping-station we may estimate the cost of our five acres at $350,000, with a probability that with favorable local conditions the expenditure would be still less. A greater number of filters would, of course, be designed to provide for increasing population, but only so many need be constructed as will meet the present requirement or that of the next two or three years, and additional filters can be added at about the same proportional cost when they are needed.

The heaviest cost for operation will be the cleaning of the filters. Estimating that 50,000,000 gallons pass each filter between scrapings, the total number of scrapings of one filter in a year will be 58, or about once a month for each of them, necessitating about 1700 days' labor. The labor for the annual deeper scraping and replacing the sand and for sand-washing can be estimated at as much more, necessitating the permanent employment of twelve men. Three men, one for each shift, will be kept always on duty to tend the gates, superintend work, and act as watchman. We may then estimate the cost of filtration for a plant of this size as follows:

COST OF OPERATION.

12 laborers, 300 days, at $2	$7,200
3 gatemen, 365 " , " $3	3,285
New sand and other supplies	1,000
Superintendence, bacterial examination of effluents, and experiments	4,000
Total cost of operation	$15,485
Interest and sinking fund on $350,000 at 6%	21,000
Total cost of filtering 365 times 8,000,000 gallons	$36,485

Equal to 1.25 cents per thousand gallons.

It is often thought that a great economy would result from filtering at a higher rate than that used in the above calculation. The following table has been prepared to show that this is not the case,

RELATIVE COST OF FILTERING AT VARIOUS RATES.

AREAS AND COSTS FOR ONE MILLION GALLONS AVERAGE YIELD.

Rate. Million Gallons per Acre Daily.	Area Available for Use at one time, allowing 25% extra for Maximum Consumption.	Reserve Area for Cleaning.* Acres.	Total Area required. Acres.	Cost at $66,000 per Acre.	Interest and Sinking Fund, 6%.	Operating Expenses.	Total Cost One Year.	Relative Cost.
0.5	2,500 acres	0.050	2.550	$168,000	$10,100	$1500	$11,600	320
1.0	1,250 "	"	1.300	86,000	5,150	"	6,650	184
1.5	0.833 "	"	0.883	58,300	3,500	"	5,000	138
2.0	0.625 "	"	0.675	44,500	2,680	"	4,180	116
2.5	0.500 "	"	0.550	36,300	2,180	"	3,680	102
2.57	0.486 "	"	0.536	35,400	2,120	"	3,620	100
3	0.417 "	"	0.467	30,800	1,850	"	3,350	93
4	0.312 "	"	0.362	23,900	1,440	"	2,940	81
5	0.250 "	"	0.300	19,800	1,190	"	2,690	74
6	0.208 "	"	0.258	17,000	1,020	"	2,520	70
7	0.179 "	"	0.229	15,100	910	"	2,410	67
8	0.156 "	"	0.206	13,600	820	"	2,320	64
9	0.139 "	"	0.189	12,500	750	"	2,250	62
10	0.125 "	"	0.175	11,600	690	"	2,190	61

* Based upon calculations for a very large plant. With daily quantities of less than 20,000,000 gallons, a larger allowance is necessary.

for even if a rate twice as great as that recommended were used, the cost would be 74 per cent as great, and with a rate four times as high the cost would still be 61 per cent of the estimate.

THE OBJECTS OF FILTRATION.

1. **Removal of** Turbidity.—The earliest filters were constructed largely, if not mainly, to remove the visible impurities from muddy river-waters, and this object remains an important one at the present time. For my part, I find as much pleasure in a bright glass of water drawn from the city tap as in a stately city hall; and a muddy fluid on the table is at least as repulsive as an ugly building. I am thus inclined purely from an æsthetic point of view to consider the filtration of turbid public water-supplies as equally important with the construction of handsome public buildings.

2. **Removal of Tastes and Odors.**—Many pond or reservoir waters which are not muddy, and which are not infected by sewage, and which cannot be objected to on sanitary grounds, are still subject at times to the growth of low forms of plants which, either in their growth or decomposition, impart to the water disagreeable tastes and odors. In such cases filters are sometimes provided to remove the unpleasant properties as well as the organisms which produce them. The filters for the Liverpool and Bradford supplies as well as those for the new supplies of Birmingham and Middlesborough in England are mainly useful in this way.

3. **Removal of the Danger from Cholera.**—The reasons for believing that cholera is caused by polluted water are entirely similar to those in the case of typhoid fever. It was no accident that the epidemic of cholera which caused the death of 3400 persons followed the temporary supply of unfiltered water by the East London Water Company in 1866, while the rest of London remained nearly free, or that the only serious outbreak of cholera in Western Europe

in 1892 was at Hamburg, which was also the only city in Germany which used raw river-water. This latter caused the sickness of 20,000 and the death of over 8000 people within a month, and an amount of suffering and financial loss, with the panics which resulted, that cannot be estimated, but that exceeded many times the cost of the filters which have since been put in operation. Hamburg had several times before suffered severely from cholera, and the removal of this danger was a leading, although not the sole, motive for the construction of filters.

How little cities supplied with pure water have to dread from cholera is shown by the experience of Altona and other suburbs of Hamburg with good water-supplies, which had but few cases of cholera not directly brought from the latter place, and by the experience of England, which maintained uninterrupted commercial intercourse with the plague-stricken city, absolutely without quarantine, and, notwithstanding a few cases which were directly imported, the disease gained no foothold in England.

I do not know of a single modern European instance where a city with a good water-supply not directly infected by sewage has suffered severely from cholera. I shall leave to others more familiar with the facts the discussion of what happened before the introduction of modern sanitary methods, as well as of the present conditions in Asia; although I believe that in these cases also there is plenty of evidence as to the part water plays in the spread of the disease.

A considerable proportion of the water-supplies of the cities of the United States are so polluted that in case cholera should gain a foothold upon our shores we have no ground for hoping for the favorable experience of the English cities rather than the plague of Hamburg in 1892.

4. **Prevention of Typhoid Fever.**—The most characteristic and uniform result of direct pollution of public water-supplies is the typhoid fever which results among the users of the water. In the English and German cities with almost uniformly good

drinking-water, typhoid fever is already nearly exterminated, and is decreasing from year to year. American cities having unpolluted water-supplies have comparatively few deaths from this cause, although the figures never go so low as in Europe, perhaps on account of the fresh cases which are always coming in from less healthy neighborhoods in ever-moving American communities. In other American cities the death-rates from typhoid fever are many times what they ought to be and what they actually are in other cities, and the rates in various places, and in the same place at different times bear in general, a close relation to the extent of the pollution of the drinking-water. The power of suitable filtration to protect a city from typhoid fever is amply shown by the very low death-rates from this cause in London, Berlin, Breslau, and large numbers of other cities drawing their raw water from sources more contaminated than those of any but the very worst American supplies.

It is obvious that if the healthfulness of water is made the main argument for filtration, it will be necessary to show a prospect of a considerable benefit to be obtained, commensurate with the outlay involved. Great engineering works will not be undertaken to save a single life or to prevent a dozen cases of fever. Filtration must be considered in the same way as a project for the abolition of grade crossings or any other public danger: its cost must be balanced against the value of the lives saved.

In making calculations of this sort the value of a life is commonly taken at $5000. When we remember that hardly any young children and but few aged people suffer from typhoid fever; that it is essentially a disease of youth and early middle age, a very great majority of deaths being of people between the ages of 15 and 50 years; and also that it affects to substantially the same extent all classes of society, the rich as well as the poor, this does not seem an excessive valuation. In Chicago in 1891, 1997 people are reported to have died from typhoid fever, representing at this rate a loss of $9,985,000. I shall not take into account the suffering and loss of time for 15,000 or more other people who were seriously sick but did not die, as the death-figures alone are ample for my present purpose. The great majority of those cases, I think I can safely say nine tenths of them, were caused either directly or indirectly by the polluted lake-water from an inlet (since abandoned) but a little way from the outlets of numerous sewers and of the Chicago River. If it is said that the other tenth would have probably occurred with ever so good a water-supply, the above estimate does not include the thousands of people who, temporarily stopping in the city, contracted the disease and went home to suffer or die, nor the thousands of others who had left their homes to find employment in the city and who returned when they found themselves sick. The deaths which resulted in this way raised the typhoid death-rates of the entire surrounding country and even of remote places. Thus I found, in investigating the water-supply of a small city over one hundred miles from Chicago, that of the apparently too large number of deaths from typhoid fever, more than one half of the cases were clearly contracted in Chicago, and half of the others originated in another city which used river-water.

The cost of filtration for the Chicago water supply cannot be told without careful estimates, but it is hardly likely that the cost would reach the $10,000,000 loss from typhoid fever experienced in one (unusually bad) year alone.

Taking the estimated cost of filtration for 100,000 people in America given above, it will be seen that if filtration resulted in saving seventy lives in a year, their value, at $5000 each, meets the estimated cost of construction, $350,000; and if only seven lives are saved, their value is sufficient to meet the operating expenses and interest on the capital invested. We can thus say roughly that filtration will prove a profitable investment when it results in a reduction of the death-rate by at least 7 per 100,000, and the saving will be sufficient to pay for the cost of construction of the filters in a single year when the reduction is at least 70 per 100,000.

The following is a list of the cities of 50,000 inhabitants and upward in the United States, with deaths from typhoid fever and the sources of their water-supplies. The deaths and populations are from the U. S. Census for 1890; the sources of the water-supplies, from the *American Water-Works Manual* for the same year. Four cities of this size—Grand Rapids, Lincoln, St. Joseph, and Des Moines—are not included in the census returns of mortality. Two cities with less than 50,000 inhabitants with exceptionally high death-rates have been included, and at the foot of the list are given corresponding data for some large European cities for 1893.

TYPHOID FEVER DEATH-RATES AND WATER-SUPPLIES OF CITIES.

City.	Population.	Deaths from Typhoid Fever. Total.	Per 100,000 living.	Water-supply.
Birmingham...	26,178	69	264	Five Mile Creek
1. Denver......	106,713	232	217	North Platte River and wells
2. Allegheny....	105,287	192	182	Allegheny River
3. Camden......	58,313	77	132	Delaware River
4. Pittsburg.....	238,617	304	127	Allegheny and Monongahela rivers
Lawrence	44,654	54	121	Merrimac River
5. Newark......	181,830	181	100	Passaic River [Ions daily
6. Charleston....	54,955	54	98	Artesian wells yielding 1,600,000 gal-
7. Washington...	230,392	200	87	Potomac River
8. Lowell......	77,696	64	82	Merrimac River
9. Jersey City...	163,003	134	82	Passaic River
10. Louisville....	161,129	122	76	Ohio River
11. Philadelphia..	1,046,964	770	74	Delaware and Schuylkill rivers
12. Chicago......	1,099,850	794	72	Lake Michigan
13. Atlanta......	65,533	47	72	South River
14. Albany.......	94,923	67	71	Hudson River
15. Wilmington..	61,431	43	70	Brandywine Creek
16. St. Paul.....	133,156	92	69	Lakes [ervoirs
17. Troy........	60,956	42	69	Hudson River and impounding res-
18. Los Angeles..	50,395	34	67	Los Angeles River and springs
19. Nashville.....	76,168	49	64	Cumberland River
20. Cleveland....	261,353	164	63	Lake Erie
21. Richmond....	81,388	50	61	James River [reservoir
22. Hartford....	53,230	32	60	Connecticut River and impounding
23. Fall River...	74,398	44	59	Watupa Lake
24. Minneapolis..	164,738	94	57	Mississippi River
25. San Francisco	298,997	166	56	Lobus Creek, Lake Merced, and
26. Indianapolis..	105,436	57	54	White River [mountain streams
27. Cincinnati....	296,908	151	51	Ohio River
28. Memphis.....	64,495	33	51	Artesian Wells
29. Reading......	58,661	29	49	Maiden Creek and Springs
30. Baltimore....	434,439	202	47	Impounding reservoir

TYPHOID FEVER DEATH-RATES AND WATER-SUPPLIES OF CITIES.

City.	Population.	Deaths from Typhoid Fever. Total.	Per 100,000 living.	Water-supply.
31. Omaha..........	140,452	63	45	Missouri River
32. Columbus......	88,150	38	43	Surface-water and wells
33. Providence.....	132,146	53	40	Pawtuxet River
34. Kansas City....	132,716	53	40	Missouri River
35. Rochester......	133,896	53	39	Hemlock and Candice lakes
36. Evansville.....	50,756	20	39	Ohio River
37. Boston.........	448,477	174	39	Impounding reservoirs
38. Toledo.........	81,434	29	36	Maumee River
39. Cambridge.....	70,028	24	34	Impounding reservoir
40. St. Louis.......	451,770	145	32	Mississippi River
41. Scranton.......	75,215	24	32	Impounding reservoir
42. Buffalo........	255,664	80	31	Niagara River
43. Milwaukee.....	204,468	61	30	Lake Michigan
44. New Haven....	81,298	22	27	Impounding reservoir
45. Worcester.....	84,655	22	26	Impounding reservoir
46. Paterson.......	78,347	20	26	Passaic River (higher up)
47. Dayton........	61,220	15	25	Wells
48. Brooklyn......	806,343	194	24	Wells, ponds, and impounding reservoirs
49. New York.....	1,515,301	348	23	Impounding reservoir
50. Syracuse......	88,143	18	20	Impounding reservoir and springs
51. New Orleans...	242,039	45	19	Mississippi River
52. Detroit........	205,876	40	19	Detroit River
53. Lynn..........	55,727	9	16	Impounding reservoir
54. Trenton.......	57,458	9	16	Delaware River
London..........	4,306,411	719	17	Filtered Thames and Lea rivers and from wells
Glasgow.........	667,883	138	20	Loch Katrine [‡ from wells
Paris............	2,424,705	609	25	Spring water
Amsterdam......	437,892	69	16	Filtered dune-water
Rotterdam.......	222,233	12	5	Filtered Maas River
Hague...........	169,828	3	2	Filtered dune-water
Berlin...........	1,714,938	161	9	Filtered Havel and Spree rivers
Hamburg........	634,878	115	18	Filtered Elbe River
Breslau..........	353,551	37	11	Filtered Oder River
Dresden.........	308,930	14	5	Ground-water
Vienna..........	1,435,931	104	7	Spring-water

Any full discussion of these data would require intimate aquaintances with the various local conditions which it is impossible to take up in detail here, but some of the leading facts cannot fail to be instructive.

Each of the places having over 100 deaths per 100,000 from typhoid fever used unfiltered river-water. Lower in the list, but still very high, Charleston, said to have been supplied only from artesian wells, had an excessive rate; but the reported water-consumption is so low as to suggest that private wells or other means of supply were in common use. Chicago and Cleveland both drew their water from lakes where they were contaminated by their own sewage. St. Paul's supply came from ponds, of which I do not know the character. With these exceptions all of the 22 cities with over 50,000 inhabitants, at the head of the list, had unfiltered river-water.

The cities supplied from impounding reservoirs as a rule had lower death rates and are at the lower end of the list, together with some cities taking their water supplies from rivers or lakes at points where they were subject to only smaller or more remote infection. Only three of the American cities in the list were reported as being supplied entirely with ground-water.

Let us assume that with present American general conditions, with a good water-supply, there may be as many as 25 deaths from typhoid fever in 100,000, due to cases resulting from the use of impure milk, the importation of cases from other towns, and other minor causes. There are eight American cities on the list coming within this limit, including New York and Brooklyn, and all of the European cities mentioned are within, and most of them much lower than, this figure. And not all the cities within the limit have such unquestionably pure supplies as to indicate that this is really the minimum.

We have every reason to expect that eventually, when the general conditions are better, rates much lower than 25 in 100,000 will be obtained. This figure is here used only as a rate which any American city with a good water-supply might reasonably expect to reach at once in spite of the new cases constantly brought in from less healthy neighborhoods.

Using this as a basis, there were in 1890 five cities each with over 50,000 inhabitants (not including the smaller cities of Lawrence and Birmingham) and an aggregate population of 690,760 which used unfiltered surface-waters, and had so many unnecessary deaths from typhoid fever that, at $5000 each, the saving due to filtration would have paid for the entire cost of the filters in the first year they were in use. Lawrence has since built a filter, and the saving of life there in the first four months that the filter was in use, at $5000 per head, was enough to pay for the filter actually built.

Following down the list, there were sixteen other cities, with an aggregate population of 3,717,560, where filtration would have paid for itself in two years or less, and still eighteen others, with an aggregate population of 3,238,617, where filtration would have saved 7 or more lives annually per 100,000, and would have more than paid for the interest and cost of operation of filters upon the basis given above.

The twelve cities lowest on the list, which used surface-waters, with an aggregrate population of 3,675,319, had so little typhoid fever that, upon the basis given, filtration would not have paid. But even in these cases, taking into account the other advantages of filtration, as well as the fact that the unavoidable death rate from typhoid fever assumed, 25 per 100,000, might actually be much reduced by an entirely safe water-supply, it is by no means certain that filtration would not have been advantageous in many of them.

It is not my purpose to make too close comparisons between the various cities on the list; some of them may have been influenced by unusual local conditions in 1890. Others have in one way or another improved their water-supplies since that date, and there are several cities in which I know the present typhoid-fever death-rates to be materially lower than those of 1890 given in the table. On the other hand, it is equally true that a number of cities, including some of the larger ones, have since had severe epidemics of typhoid fever which have given very much higher rates than those for 1890.

These fluctuations would change the order of cities in the list from year to year; they would not change the general facts, which are as true to-day as they were in 1890. Nearly all of the great cities of the United States are supplied with unfiltered surface-waters, and a great majority of the waters are taken from rivers and lakes at points where they are polluted by sewage. The death-rates from typhoid fever in those cities, whether they are compared with better supplied cities of this country, or with European cities, are enormously high.

Such rates were formerly common in European cities, but they have disappeared with better sanitary conditions. The introduction of filters has often worked marvellous changes in Europe, and in Lawrence the improvement in the city's health with filtered water was prompt and unquestionable. There is every reason to believe that the general introduction of better water in American cities will work corresponding revolutions; and looking at it from a merely money standpoint, the value of the lives and the saving of the expenses of sickness will pay handsomely when compared with the cost of good water.

WHAT WATERS REQUIRE FILTRATION?

From the nature of the case a satisfactory general answer to this question cannot be given, but a few suggestions may be useful.

In the first place, ground-waters obviously do not require filtration: they have already in most cases been thoroughly filtered in the ground through which they have passed, and in the exceptional cases, as, for instance, an artesian well drawing water through fissures in a ledge from a polluted origin, a new supply will generally be chosen rather than to attempt to improve so doubtful a raw material.

River-waters should be filtered. It cannot be asserted that there are no rivers in montainous districts in which the water is at once clear and free from pollution, and suitable in its natural state for water-supply; but if so, they are not common, least of all in the regions where water-supplies are usually required. The use of river-waters in their natural state or after sedimentation only, drawn from such rivers as the Merrimac, Hudson, Potomac, Delaware, Schuylkill, Ohio, and Mississippi, is a filthy as well as an unhealthy practice, which ought to be abandoned.

The question is more difficult in the case of supplies drawn from lakes or storage reservoirs. Many such supplies are grossly polluted and should be either abandoned or filtered. Others are subject to algæ growths, or are muddy, and would be much improved by filtration. Still others are drawn either from unpolluted water-sheds, or the pollution is so greatly diluted and reduced by storage that no known disadvantage results from their use.

In measuring the effects of the pollution of water-supplies, the typhoid-fever death-rate is a most important aid. Not that typhoid fever is the sole evil resulting from polluted water, but because it is also a very useful index of other evils for which corresponding statistics cannot be obtained, as, for instance, the causation of diarrhœal diseases or the danger from invasion by cholera.

I think we shall not go far wrong at the start to confine our attention to those cities where there are over 25 deaths from typhoid fever per 100,000 of population. This will at once throw out of consideration a large number of relatively good supplies, including those of New York and Brooklyn. It is not my idea that none of these supplies cause disease. Many of them, as for instance that of New York, are known to receive sewage, and it is an interesting question worthy of most careful study whether there are cases of sickness resulting from this pollution. The point that I wish to make now is simply that in those cases the death-rate itself is evidence that, with existing conditions of dilution and storage, the resulting damage of which we have knowledge is not great enough to justify the expense involved by filtration.

In this connection it should not be forgotten that, especially with very small watersheds, there may be a danger as distinct from present damage which requires consideration. Thus a single house or groups of houses draining into a supply may not appreciably affect it for years, until an outbreak of fever on the water-shed results in infecting the water with the germs of disease and in an epidemic in the city below. This danger decreases with increasing size of the water-shed and volume of the water with which any such pollution would be mixed, and also with the population draining into the water, as there is a probability that the amount of infection continually added from a considerable town will not be subject to as violent fluctuation as that from only a few houses.

Thus in Plymouth, Pa., in 1885, there were 1104 cases of typhoid fever and 114 deaths among a population of 8000, as the result of the discharge of the dejecta from a single typhoid patient into the water of a relatively small impounding reservoir. The cost of this epidemic was calculated with unusual care. The care of the sick cost in cash $67,100.17, and the loss of wages for those who recovered amounted to $30,020.08. The 114 persons who died were earning before their sickness at the rate of $18,419.52 annually.

Such an outbreak would hardly be possible with the Croton water-shed of the New York water-supply, on account of the great dilution and delay in the reservoirs, but it must be guarded against in small supplies.

Of the cities having more than 25 deaths per 100,000 from typhoid fever, some will no doubt be found where milk epidemics or other special circumstances were the cause; but I believe in a majority of them, and in nearly all cases where the rate is year after year considerably above that figure, the cause will be found in the water-supply. Investigation should be made of this point; and if the water is not at fault, the responsibility should be located. If the water is guilty, it should be either purified or a new supply obtained.

CHAPTER X.

WATER-SUPPLY AND DISEASE—CONCLUSIONS.

The fæces from a man contain on an average perhaps 1,000,000,000 bacteria per gram,* most of them being the normal bacilli of the intestines, *Bacillus coli communis*. Assuming that a man discharges 200 grams or about 7 ounces of fæces daily, this would give 200,000,000,000 bacteria discharged daily per person. The number of bacteria actually found in American sewage is usually higher, often double this number per person; but there are other sources of bacteria in sewage, and in addition growths or the reverse may take place in the sewers, according to circumstances.

This number of bacteria in sewage is so enormously large that the addition of the sewage from a village or city to even a large river is capable of affecting its entire bacterial composition. Thus taking the population of Lowell in 1892 at 85,000, and the average daily flow of the Merrimac at 6000 cubic feet per second, and assuming that 200,000,000,000 bacteria are discharged daily in the sewage from each person, they would increase the number in the river by 1160 per cubic centimeter, or about 300,000 in an ordinary glass of water. The average number found in the water eight miles below, at the intake of the Lawrence water-works, was more than six times as great as this, due in part to the sewage of other cities higher up.

There is every reason to believe that the bulk of these bacteria

---

* This number was the result of numerous counts made from fæces from persons suffering with typhoid fever in the Lawrence City Hospital in 1891 and 1892. Mr. G. W. Fuller, now in charge of the Lawrence Experiment Station, has kindly made some further investigation of fæces from healthy people in which the numbers have been considerably lower, usually less than 200,000,000, per gram and sometimes as low as 10,000,000 per gram.

were harmless to the people of Lawrence, who drank them; but some of them were not. Fæces of people suffering from typhoid fever contain the germs of that disease. What proportion of the total number of bacteria in such fæces are injurious is not known; but assuming that one fourth only of the total number are typhoid germs, and supposing the fæces of one man to be evenly mixed with the whole daily average flow of the river, it would put one typhoid germ into every glass of water at the Lawrence intake, and at low water several times as many proportionately would be added. This gives some conception of the dilution required to make a polluted water safe.

One often hears of the growth of disease-germs in water, but as far as the northern United States and Europe are concerned there is no evidence whatever that this ever takes place. There are harmless forms of bacteria which are capable of growing upon less food than the disease-germs require and they often multiply in badly-polluted waters. Typhoid-fever germs live for a longer or shorter period, and finally die without growth. The few laboratory experiments which have seemed to show an increase of typhoid germs in water have been made under conditions so widely different from those of natural watercourses that they have no value.*

The proportionate number of cases of typhoid fever among the users of a polluted water varies with the number of typhoid germs in the water. Excessive pollution causes severe epidemics or continued high death-rates according as the infection is continued or

---

* These experiments, so far as they have come to the notice of the author, have been made with water sterilized by heating, usually in small tubes stoppered with cotton-wool or other organic matter. In this case the water, no matter how carefully purified in the first place, becomes an infusion of organic matters capable of supporting bacterial growths, and not at all to be compared to natural waters.

In experiments often repeated under my direction, carefully distilled water in bottles, *most scrupulously clean*, with glass stoppers, and protected from dust, but *not sterilized*, has uniformly refused to support bacterial growths even when cautiously seeded at the start, and the same is usually true of pure natural waters. Some further experiments showed hardly any bacterial growth even of the most hardy water bacteria in a solution 1 part of peptone in 1,000,000,000 parts of distilled water, and solutions ten times as strong only gave moderate growths.

intermittent. Slight infection causes relatively few cases of fever. Pittsburg and Allegheny, taking their water-supplies from below the outlets of some of their own sewers, have suffered severely (103.2 and 127.4 deaths from typhoid fever annually per 100,000, respectively, from 1888 to 1892). Wheeling, W. Va., with similar conditions in 1890, was even worse, a death rate of 345 per 100,000 from this cause being reported, while Albany had only comparatively mild epidemics from the less directly and grossly polluted Hudson. Lawrence and Lowell, taking their water from the Merrimac, both had for many years continued excessive rates, increasing gradually with increasing pollution; and the city having the most polluted source had the higher rate.

In Berlin and Altona, in winter, with open filters, epidemics of typhoid fever followed decreased efficiency of filtration, but the epidemics were often so mild that they would have entirely escaped observation under present American conditions. Chicago has for years suffered from typhoid fever, and the rate has fluctuated, as far as reliable information can be obtained with the fluctuations in the pollution of the lake water. An unusual discharge of the Chicago River results in a higher death-rate. Abandoning the shore inlet near the mouth of the Chicago River in 1892, resulted in the following year in a reduction of 60 per cent in the typhoid fever death-rate.* This reduction shows, not that the present intakes are safe, but simply that they are less polluted than the old ones to an extent measured by the reduction in the death-rate.

It is not supposed that in an epidemic of typhoid fever caused by polluted water every single person contracts the disease directly by drinking the water. On the contrary, typhoid fever is often communicated in other ways. If we have in the first place a thousand cases in a city caused directly by the water, they will be followed by a large number of other cases resulting directly from the pres-

---

* The Water-supply of Chicago: Its Source and Sanitary Aspects. By Arthur R. Reynolds, M.D., Commissioner of Health of Chicago, and Allen Hazen. *American Public Health Association*, 1893. Page 146.

ence in the city of the first thousand cases. The conditions favoring this spread may vary in different wards, resulting in considerable local variations in the death-rates. Some persons also will suffer who did not drink any tap-water. These facts, always noted in epidemics, afford no ground for refusing to believe, in the presence of direct evidence, that the water was the cause of the fever. These additional cases are the indirect if not the direct result of the water. The broad fact that cities with polluted water-supplies as a rule have high typhoid-fever death-rates, and cities with good water-supplies do not (except in the occasional cases of milk epidemics, or where they are overrun by cases contracted in neighboring cities with bad water, as is the case with some of Chicago's suburbs), is at once the best evidence of the damage from bad water and measure of its extent.

The conditions which remove or destroy the sewage bacteria in a water tend to make it safe. The most important of them are: (1) dilution (2) time, allowing the bacteria to die (sunlight may aid in this process, although effective sunshine cannot reach the lower layer of turbid waters or through ice); (3) sedimentation, allowing them to go to the bottom, where they eventually die; and (4) natural or artificial filtration. In rivers, distance is mainly useful in affording time, and also, under some conditions, in allowing opportunities for sedimentation. Thus a distance of 500 miles requires a week for water travelling three miles an hour to pass, and will allow very important changes to take place. The old theory that water purifies itself in running a certain distance has no adequate foundation as far as bacteria are concerned. Some purification takes place with the time involved in the passage, but its extent has been greatly overestimated.

The time required for the bacteria to die simply from natural causes is considerable; certainly not less than three or four weeks can be depended upon with any confidence. In storage reservoirs this action is often considerable, and it is for this reason that American water-supplies from large storage reservoirs are, as a rule, much more healthy than those drawn from rivers or polluted lakes, even when the sources of the former are somewhat polluted. The water-supplies of New York and Boston may be cited as examples. In many other water-works operations the entire time from the pollution to the consumption of the water is but a few days or even less, and time does not materially improve water in this period.

Sedimentation removes bacteria only slowly, as might be expected from their exceedingly small size; and in addition their specific gravity probably is but slightly greater than that of water. The Lawrence reservoir, holding from 10 to 14 days' supply, effected, by the combined effect of time and sedimentation, a reduction of 90 per cent of the bacteria in the raw water. In spite of this the city suffered severely and continuously from fever. It would probably have suffered even more, however, had it not been for this reduction. Nothing is known of the removal of bacteria by sedimentation from flowing rivers, but, considering the slowness with which the process takes place in standing water, it is evident that we cannot hope for very much in streams, and especially rapid streams, where the opportunities for sedimentation are still less favorable.

Filtration as practiced in Europe removes promptly and certainly a very large proportion of the bacteria—probably, under all proper conditions, over 99 per cent, and is thus much more effective in purification than even weeks of storage or long flows in rivers. The places using filtered water have, in general, extremely low death-rates from typhoid fever. The fever which has occurred at a few places drawing their raw water from greatly polluted sources has resulted from improper conditions which can be avoided, and affords no ground for doubt of the efficiency of properly conducted filtration.

The same confidence cannot be expressed in regard to the mechanical filters often used in America for filtering at much higher rates than those used in European filters. There is good evidence that so far as they do not use alum they cannot even approximate to the result obtained by slow sand filtration. With the use of alum in sufficient quantity better bacterial results are obtained, but the evidence is not conclusive either that a satisfactory purification can or cannot be obtained with them. The use of such filters will not, under any circumstances, add infection to the water, and will undoubtedly remove some of the bacteria. Between no filtration and mechanical filtration of an infected water I should by all means prefer the latter. I should, however, insist upon thorough and continued chemical and bacterial examinations of the effluent. If it was found that bacteria were passing too freely, the examinations would show the danger. And in case of an epidemic of typhoid fever or cholera the experience gained, although dear, would be extremely valuable and conclusive.

The main point is that disease-germs shall not be present in our drinking-water. If they can be kept out in the first place at reasonable expense, that is the thing to do. Innocence is better than repentance. If they cannot be kept out, we must take them out afterwards; it does not matter much how this is done, so long as the work is thorough. Sedimentation and storage may accomplish much, but their action is too slow and often uncertain. Filtration properly carried out removes bacteria promptly and thoroughly and at a reasonable expense.

APPENDICES.

APPENDIX I.

RULES OF THE GERMAN GOVERNMENT IN REGARD TO THE FILTRATION OF SURFACE-WATERS USED FOR PUBLIC WATER-SUPPLIES.

Rules somewhat similar to those of which a translation is given below were first issued by the Imperial Board of Health in 1892. These rules were regarded as unnecessarily rigid, and a petition was presented to the government signed by 37 water-works engineers and directors requesting a revision.* As a result a conference was organized consisting of 14 members.† Köhler presided, and Koch, Gaffsky, Werner, Günther, and Reincke represented the Imperial Board of Health. The bacteriologists were represented by Flügge, Wolffhügel, and Fränkel, while Beer, Fischer, Lindley, Meyer, and Piefke were the engineer members.

This conference prepared the 17 articles given below in the first days of January, 1894. A little later the first 16 articles were issued to all German local authorities, signed by Bosse, minister of the "Geistlichen," and Haase, minister of the interior, and they are considered as binding upon all water-works using surface-water. The bacterial examinations were commenced April 1, 1894, by most of the cities which had not previously had them.

---

* *Journal für Gas- u. Wasserversorgung*, 1893, 694.
† *Journal für Gas- u. Wasserversorgung*, 1894, 185.

Although the articles do not deal with rate of filtration, or the precautions against snow and ice, they have a very great interest both because they are an official expression, and on account of the personal standing of the men who prepared them.

§ 1. In judging of the quality of a filtered surface-water the following points should be especially observed:

a. The operation of a filter is to be regarded as satisfactory when the filtrate contains the smallest possible number of bacteria, not exceeding the number which practical experience has shown to be attainable with good filtration at the works in question. In those cases where there are no previous records showing the possibilities of the works and the influence of the local conditions, especially the character of the raw water, and until such information is obtained, it is to be taken as the rule that a satisfactory filtration will never yield an effluent with more than about 100 bacteria per cubic centimeter.

b. The filtrate must be as clear as possible, and, in regard to color, taste, temperature, and chemical composition, must be no worse than the raw water.

§ 2. To allow a complete and constant control of the bacterial efficiency of filtration, the filtrate from each single filter must be examined daily. Any sudden increase in the number of bacteria should cause a suspicion of some unusual disturbance in the filter, and should make the superintendent more attentive to the possible causes of it.

§ 3. Filters must be so constructed that samples of the effluent from any one of them can be taken at any desired time for the bacteriological examination mentioned in § 1.

§ 4. In order to secure uniformity of method, the following is recommended as the standard method for bacterial examination:

The nutrient medium consists of 10 per cent meat extract gelatine with peptone, 10 cc. of which is used for each experiment. Two samples of the water under examination are to be taken, one of 1 cc. and one of ½ cc. The gelatine is melted at a temperature of 30° to 35° C., and mixed with the water as thoroughly as possible in the test-tube by tipping back and forth, and is then poured upon a sterile glass plate. The plates are put under a bell-jar which stands upon a piece of blotting-paper saturated with water, and in a room in which the temperature is about 20° C.

The resulting colonies are counted after 48 hours, and with the aid of a lens.

If the temperature of the room in which the plates are kept is lower than the above, the development of the colonies is slower, and the counting must be correspondingly postponed.

If the number of colonies in 1 cc. of the water is greater than about 100, the counting must be done with the help of the Wolffhügel's apparatus.

§ 5. The person entrusted with the carrying-out of the bacterial examinations must present a certificate that he possesses the necessary qualifications, and wherever possible he shall be a regular employé of the water-works.

§ 6. When the effluent from a filter does not correspond to the hygienic requirements it must not be used, unless the cause of the unsatisfactory work has already been removed during the period covered by the bacterial examinations.

In case a filter for more than a very short time yields a poor effluent, it is to be put out of service until the cause of the trouble is found and corrected.

It is, however, recognized from past experience that sometimes unavoidable conditions (high water, etc.) make it impossible, from an engineering standpoint, to secure an effluent of the quality stated in § 1. In such cases it will be necessary to get along with a poorer quality of water; but at the same time, if the conditions demand it (outbreak of an epidemic, etc.), a suitable notice should be issued.

§ 7. Every single filter must be so built that, when an inferior effluent results, which does not conform to the requirements, it can be disconnected from the pure-water pipes and the filtrate allowed to be wasted, as mentioned in § 6. This wasting should in general take place, so far as the arrangement of the works will permit it:

(1) Immediately after scraping a filter; and
(2) After replacing the sand to the original depth.

The superintendent must himself judge, from previous experience with the continual bacterial examinations, whether it is necessary to waste the water after these operations, and, if so, how long a time will probably elapse before the water reaches the standard purity.

§ 8. The best sand-filtration requires a liberal area of filter-surface, allowing plenty of reserve, to secure, under all local conditions, a moderate rate of filtration adapted to the character of the raw water.

§ 9. Every single filter shall be independently regulated, and the rate of filtration, loss of head, and character of the effluent shall be known. Also each filter shall, by itself, be capable of being completely emptied, and, after scraping, of having filtered water introduced from below until the sand is filled to the surface.

§ 10. The velocity of filtration in each single filter shall be capable of being arranged to give the most favorable results, and shall be as regular as possible, quite free from sudden changes or interruptions. On this account reservoirs must be provided large enough to balance the hourly fluctuation in the consumption of water.

§ 11. The filters shall be so arranged that their working shall not be influenced by the fluctuating level of the water in the filtered-water reservoir or pump-well.

§ 12. The loss of head shall not be allowed to become so great as to cause a breaking through of the upper layer on the surface of the filter. The limit to which the loss of head can be allowed to go without damage is to be determined for each works by bacterial examinations.

§ 13. Filters shall be constructed throughout in such a way as to insure the equal action of every part of their area.

§ 14. The sides and bottoms of filters must be made water-tight, and special pains must be taken to avoid the danger of passages or loose places through which the unfiltered water on the filter might find its way to the filtered-water channels. To this end special pains should be taken to make and keep the ventilators for the filtered-water channels absolutely tight.

§ 15. The thickness of the sand-layer shall be so great that under no circumstances shall it be reduced by scraping to less than 30 cm. (= 12 inches), and it is desirable, so far as local conditions allow, to increase this minimum limit.

Special attention must be given to the upper layer of sand, which must be arranged and continually kept in the condition most favorable for filtration. For this reason it is desirable that, after a filter has been reduced in thickness by scraping and is about to be refilled, the sand below the surface, as far as it is discolored, should be removed before bringing on the new sand.

§ 16. Every city in the German empire using sand-filtered water is requested to make a quarterly report of their working results, especially of the bacterial character of the water before and after filtration, to the Imperial Board of Health (Kaiserlichen Gesundheitsamt), which will keep itself in communication with the commission chosen by the water-works engineers in regard to these questions; and it is believed that after such statistical information is obtained for a period of about two years some farther judgments can be reached.

§ 17. The question as to the establishment of a permanent inspection of public water-works, and, if so, under what conditions, can be best answered after the receipt of the information indicated in § 16.

APPENDIX II.

EXTRACTS FROM "BERICHT DES MEDICINAL-INSPECTORATS DES HAMBURGISCHEN STAATES FÜR DAS JAHR 1892."

THE following are translations from Dr. Reincke's most valuable report upon the vital statistics of Hamburg for 1892. I much regret that I am unable to reproduce in full the very complete and instructive tables and diagrams which accompany the report.

Diarrhœa and Cholera Infantum (page 10). "It is usually assumed that the increase of diarrhœal diseases in summer is to be explained by the high temperature, especially by the action of the heat upon the principal food of infants—milk. Our observations, however, indicate that a deeper cause must be sought." (Tables and diagrams of deaths from cholera infantum by months for Hamburg and for Altona with the mean temperatures, 1871–1892.)

"From these it appears that the highest monthly mortality of each year in Hamburg occurred 7 times in July, 13 times in August, and 3 times in September, and substantially the same in Altona. If one compares the corresponding temperatures, it is found that in the three years 1886, 1891, and 1892, with high September mortalities, especially the first two of them, had their maximum temperature much earlier, in fact earlier than usual. Throughout, the correspondence between deaths and temperatures is not well marked. Repeated high temperatures in May and June have never been followed by a notable amount of cholera infantum, although such periods have lasted for a considerable time. For example, toward the end of May, 1892, for a long time the temperature was higher than in the following August, when the cholera infantum appeared.

"The following observations are still more interesting. As is seen from the diagram, in addition to the annual rise in summer there is also a smaller increase in the winter, which is especially marked in Altona. In 1892 this winter outbreak was greater than the summer one, and nearly as great in 1880 and in 1888. The few years when this winter increase was not marked, 1876-7, 1877-8, 1881-2, 1883-4, were warm winters in which the mean temperature did not go below the freezing-point. It is also to be noted that the time of this winter outbreak is much more variable than that of the summer one. In 1887 the greatest mortality was in November; in 1889 in February; in other years in December or January, and in Altona, in 1886 and 1888, in March, which is sufficient evidence that it was not the result of Christmas festivities.

"Farther, the winter diarrhœa of Hamburg and of Altona are not parallel as is the case in summer. In Hamburg the greatest mortality generally comes before New Year's: in Altona one to two months later.

"In Bockendahl's Generalbericht über das öffentliche Gesundheitswesen der Provinz Schleswig-Holstein für das Jahr 1870, page 10, we read: 'Yet more remarkable was an epidemic of cholera infantum in Altona in February which proved fatal to 43 children. These cases were distributed in every part of the city, and could not be explained by the health officer until he ascertained that the water company had supplied unfiltered water to the city. This occurred for a few days only in January, and was the only time in the whole year that unfiltered Elbe water was delivered. However little reason there may be to believe that there was a connection between these circumstances, future interruptions of the service of filtered water should be most critically watched, as only in this way can reliable conclusions be reached. Without attempting to draw any scientific conclusions from the fact, I cannot do less than record that, prior to the outbreak of cholera on August 20, 1871, unfiltered together with filtered water had been supplied to the city August 11 to 18. The action of the authorities was then justified when they forbade in future the supply of unfiltered water except in cases of most urgent necessity, as in case of general conflagration; and in such a case, or in case of interruption due to broken pipes, that the public should be suitably warned.'

"The author of this paragraph, Dr. Kraus, became later the health officer of Hamburg, and in an opinion written by him in 1874, and now before me, he most earnestly urged the adoption of sand-filtration in Hamburg, and cites the above observations in support of his position. In the annual report of vital statistics of Hamburg for 1875 he says that it is quite possible that the addition of unfiltered Elbe water to milk is the cause of the high mortality from cholera infantum, as compared with London, and this idea was often afterward expressed by him. Since then so much evidence has accumulated that his view may fairly be considered proved.

"For the information of readers not familiar with local conditions, a mention of the sources of the water-supplies up to the present time used by Hamburg and Altona will be useful. Both cities take their entire water-supplies from the Elbe—Altona from a point about 7 miles below the discharge of the sewage of both cities, Hamburg from about 7 miles above. The raw water at Altona is thus polluted by the sewage from the population of both cities, having now together over 700,000 inhabitants, and contains in general 20,000 to 40,000 or more bacteria per cubic centimeter. The raw water of Hamburg has, however, according to the time of year and tide, from 200 to 5000, but here also occasionally much higher numbers are obtained when the ebb tide carries sewage up to the intake. How often this takes place is not accurately known, but most frequently in summer when the river is low, more rarely in winter and in times of flood. Recent bacterial examinations show that it occurs much more frequently than was formerly assumed from float experiments. This water is pumped directly to the city raw, while that for Altona is carefully filtered.

"Years ago I expressed the opinion that the repeated typhoid epidemics in Altona stood in direct connection with disturbances of the action of the filters by frost, which result in the supply of insufficiently purified water. Wallichs in Altona has also come to this conclusion as a result of extended observation, and recently Robert Koch has explained the little winter epidemic of cholera in Altona in the same way, thus supporting our theory. When open filters are cleaned in cold, frosty weather the bacteria in the water are not sufficiently held back by the filters. Such disturbances of filtration not only preceded the explosive epidemics of typhoid fever of 1886, 1887, 1888, 1891, and 1892, and the cholera outbreaks of 1871 and 1893, but also the winter outbreaks of cholera infantum which have been so often repeated. It cannot be doubted that these phenomena bear the relation to each other of cause and effect. It is thus explained why in the warm winters no such outbreaks have taken place, and also why the cholera infantum in winter is not parallel in Hamburg and Altona.

. . . .

"A farther support of this idea is furnished by Berlin, where in the same way frost has repeatedly interfered with filtration. In the following table are shown the deaths from diarrhœa and cholera infantum for a few winter periods having unusual increases in mortality in comparison with the bacteria in the water-supply." (These tables show that in March, 1886, March, 1888, February—March, 1889, and February, 1891, high numbers of bacteria resulted from frost disturbance at the Stralau works, and in every case they were followed by greatly increased death-rates from diarrhœal diseases. —A. H.)

"No one who sees this exhibition can doubt that here also the supply of inadequately purified water has every time cost the lives of many children." (100 to 400 or more each time.—A. H.) "Even more conclusive is the evidence, published by the Berlin Health Office, that this increase was confined to those parts of the city supplied from Stralau" (with open filters.—A. H.), "and that the parts supplied from the better Tegel works took no part in the outbreaks, which was exactly the case with the well-known typhoid epidemic of February and March, 1889. . . . It was also found that those children nursed by their mothers or by wet-nurses did not suffer, but only those fed on the milk of animals or other substitutes, and which in any case were mixed with more or less water."

Under **Cholera**, page 28, he says: "The revised statistics here given differ slightly from preliminary figures previously issued and widely published." (The full tables, which cannot be here reproduced, show 16,956 cases and 8605 deaths. 8146 of the deaths occurred in the month ending September 21. Of these, 1799 were under 5 years old; 776 were 5 to 15; 744, 15 to 25; 3520, 25 to 50; 1369, 50 to 70; and 397 over 70 or of unknown age. The bulk of the cases were thus among mature people, children, except very young children, suffering the least severely of any age class.)

"The epidemic began on August 16, in the port where earlier outbreaks have also had their origin. The original source of the infection has not been ascertained with certainty, but was probably from one of two sources. Either it came from certain Jews, just arrived from cholera-stricken Russia, who were encamped in large numbers near the American pier, or the infection came from Havre, where cholera had been present from the middle of July. Perhaps the germs came in ships in water-ballast which was discharged at Hamburg, which is so much more probable, as the sewage of Havre is discharged directly into the docks.

"It is remarkable that in Altona, compared to the total number of cases, very few children had cholera, while in the epidemic of 1871 the children suffered severely. This may be explained by supposing that the cholera of 1892 in Altona was not introduced by water, but by other means of infection. . . .

"It is well known that the drinking-water (of Hamburg) is supposed to have been from the first the carrier of the cholera-germs. In support of this view the following points are especially to be noted:

"1. The explosive rapidity of attack. The often-compared epidemic in Munich in 1854, which could not have come from the water is characteristically different in that its rise was much slower and was followed by a gradual decline. In Hamburg, with six times as large a population, the height of the epidemic was reached August 27, only 12 days after the first cases of sickness, while in Munich 25 days were required. In Hamburg also the bulk of the cases were confined to 12 days, from August 25 to September 5, while in Munich the time was twice as long.

"2. The exact limit of the epidemic to the political boundary between Hamburg and Altona and Wandsbeck, which also agrees with the boundary between the respective water-supplies, while other differences were entirely absent. Hamburg had for 1000 inhabitants 26.31 cases and 13.39 deaths, but Altona only 3.81 cases and 2.13 deaths, and Wandsbeck 3.06 cases and 2.09 deaths. . .

"3. The old experience of cholera in fresh-water ports, and the analogy of many earlier epidemics. In this connection the above-mentioned epidemic of 1871 in Altona has a special interest, even though some of the conclusions of Bockendahl's in his report of 1871 are open to objection. First there were 3 deaths August 3, which were not at once followed by others. Then unfiltered Elbe water was supplied August 11 to 18. On the 19th an outbreak of cholera extended to all parts of the city, which reached its height August 25 and 26, and afterwards gradually decreased. In all 105 persons died of cholera and 186 (179 of them children) of diarrhœa. In Hamburg, four times as large, only 141 persons died of cholera at this time, thus proportionately a smaller number. The conditions were then the reverse of those of 1892, an infection of the Altona water and a comparative immunity in Hamburg.

"It is objected that the cholera-germs were not found in the water in 1892. To my knowledge they were first looked for, and then with imperfect methods, in the second half of September. In the after-epidemics at Altona, they were found in the river-water by R. Koch by the use of better methods.

"It is quite evident that the germs were also distributed by other methods than by the city water, especially by dock-laborers who became infected while at their work and thus set up little secondary epidemics where they went or lived. . . . These laborers and sailors, especially on the smaller river-boats, had an enormously greater proportionate amount of cholera than others. . . . These laborers do not live exclusively near the water, but to a measure in all parts of the city." (And in Altona and Wandsbeck.—A. H.)

"Altona had 5 deaths from cholera December 25 to January 4, and 19 January 23 to February 11, and no more. As noted above, this is attributed to the water-supply, and to defective filtration in presence of frost. . . .

"The cholera could never have reached the proportion which it did, had the improvements in the drinking-water been earlier completed."

Further accounts of the water-supplies of Altona and of Hamburg and of the new filtration works at the latter city are given in Appendices VII and VIII.

APPENDIX III.

METHODS OF SAND-ANALYSIS.

(From the Annual Report of the Massachusetts State Board of Health for 1892.)

A KNOWLEDGE of the sizes of the sand-grains forms the basis of many of the computations. This information is obtained by means of mechanical analyses. The sand sample is separated into portions having grains of definite sizes, and from the weight of the several portions the relative quantities of grains of any size can be computed.

Collection of Samples.—In shipping and handling, samples of sand are best kept in their natural moist condition, as there is then no tendency to separation into portions of unequal-sized grains. Under no circumstances should different materials be mixed in the same sample. If the material under examination is not homogeneous, samples of each grade should be taken in separate bottles, with proper notes in regard to location, quantity, etc. Eight-ounce wide-necked bottles are most convenient for sand samples, but with gravels a larger quantity is often required. Duplicate samples for comparison after obtaining the results of analyses are often useful.

Separation into Portions having Grains of Definite Sizes.—Three methods are employed for particles of different sizes—hand-picking for the stones, sieves for the sands, and water elutriation for the extremely fine particles. Ignition, or determination of albuminoid ammonia, might be added for determining the quantity of organic matter, which, as a matter of convenience, is assumed to consist of particles less than 0.01 millimeter in diameter.

The method of hand-picking is ordinarily applied only to particles which remain on a sieve two meshes to an inch. The stones of this size are spread out so that all are in sight, and a definite number of the largest are selected and weighed. The diameter is calculated from the average weight by the method to be described, while the percentage is reckoned from the total weight. Another set of the largest remaining stones is then picked out and weighed as before, and so on until the sample is exhausted. With a little practice the eye enables one to pick out the largest stones quite accurately.

With smaller particles this process becomes too laborious, on account of the large number of particles, and sieves are therefore used instead. The sand for sifting must be entirely free from moisture, and is ordinarily dried in an oven at a temperature somewhat above the boiling-point. The quantity taken for analysis should rarely exceed 100–200 grams. The sieves are made from carefully-selected brass-wire gauze, having, as nearly as possible, square and even-sized meshes. The frames are of metal, fitting into each other so that several sieves can be used at once without loss of material. It is a great convenience to have a mechanical shaker, which will take a series of sieves and give them a uniform and sufficient shaking in a short time; but without this good results can be obtained by hand-shaking. A series which has proved very satisfactory has sieves with approximately 2, 4, 6, 10, 20, 40, 70, 100, 140, and 200 meshes to an inch; but the exact numbers are of no consequence, as the actual sizes of the particles are relied upon, and not the number of meshes to an inch.

It can be easily shown by experiment that when a mixed sand is shaken upon a sieve the smaller particles pass first, and as the shaking is continued larger and larger particles pass, until the limit is reached when almost nothing will pass. The last and largest particles passing are collected and measured, and they represent the separation of that sieve. The size of separation of a sieve bears a tolerably definite relation to the size of the mesh, but the relation is not to be depended upon, owing to the irregularities in the meshes and also to the fact that the finer sieves are woven on a different pattern from the coarser ones, and the particles passing the finer sieves are somewhat larger in proportion to the mesh than is the case with the coarser sieves. For these reasons the sizes of the sand-grains are determined by actual measurements, regardless of the size of the mesh of the sieve.

It has not been found practicable to extend the sieve-separations to particles below 0.10 millimeter in diameter (corresponding to a sieve with about 200 meshes to an inch), and for such particles elutriation is used. The portion passing the finest sieve contains the greater part of the organic matter of the sample, with the exception of roots and other large undecomposed matters, and it is usually best to remove this organic matter by ignition at the lowest possible heat before proceeding to the water-separations. The loss in weight is regarded as organic matter, and calculated as below 0.01 millimeter in diameter. In case the mineral matter is decomposed by the necessary heat, the ignition must be omitted, and an approximate equivalent can be obtained by multiplying the albuminoid ammonia of the sample by 50.* In this case it is necessary to deduct an equivalent amount from the other fine portions, as otherwise the analyses when expressed in percentages would add up to more than one hundred.

Five grams of the ignited fine particles are put in a beaker 90 millimeters high and holding about 230 cubic centimeters. The beaker is then nearly filled with distilled water at a temperature of 20° C., and thoroughly mixed by blowing into it air through a glass tube. A larger quantity of sand than 5 grams will not settle uniformly in the quantity of water given, but less can be used if desired. The rapidity of settlement depends upon the temperature of the water, so that it is quite important that no material variation in temperature should occur. The mixed sand and water is allowed

---

* The method of making this determination was given in the *American Chemical Journal*, vol. 12, p. 427.

to stand for fifteen seconds, when most of the supernatant liquid, carrying with it the greater part of the particles less than 0.08 millimeter, is rapidly decanted into a suitable vessel, and the remaining sand is again mixed with an equal amount of fresh water, which is again poured off after fifteen seconds, carrying with it most of the remaining fine particles. This process is once more repeated, after which the remaining sand is allowed to drain, and is then dried and weighed, and calculated as above 0.08 millimeter in diameter. The finer decanted sand will have sufficiently settled in a few minutes, and the coarser parts at the bottom are washed back into the beaker and treated with water exactly as before, except that one minute interval is now allowed for settling. The sand remaining is calculated as above 0.04 millimeter, and the portion below 0.04 is estimated by difference, as its direct determination is very tedious, and no more accurate than the estimation by difference when sufficient care is used.

Determination of the Sizes of the Sand-grains.—The sizes of the sand-grains can be determined in either of two ways—from the weight of the particles or from micrometer measurements. For convenience the size of each particle is considered to be the diameter of a sphere of equal volume. When the weight and specific gravity of a particle are known, the diameter can be readily calculated. The volume of a sphere is $\frac{1}{6}\pi d^3$, and is also equal to the weight divided by the specific gravity. With the Lawrence materials the specific gravity is uniformly 2.65 within very narrow limits, and we have $\frac{w}{2.65} = \frac{1}{6}\pi d^3$. Solving for d we obtain the formula $d = .9\sqrt[3]{w}$, where d is the diameter of a particle in millimeters and w its weight in milligrams. As the average weight of particles, when not too small, can be determined with precision, this method is very accurate, and altogether the most satisfactory for particles above 0.10 millimeter; that is, for all sieve separations. For the finer particles the method is inapplicable, on account of the vast number of particles to be counted in the smallest portion which can be accurately weighed, and in these cases the sizes are determined by micrometer measurements. As the sand-grains are not spherical or even regular in shape, considerable care is required to ascertain the true mean diameter. The most accurate method is to measure the long diameter and the middle diameter at right angles to it, as seen by a microscope. The short diameter is obtained by a micrometer screw, focussing first upon the glass upon which the particle rests and then upon the highest point to be found. The mean diameter is then the cube root of the product of the three observed diameters. The middle diameter is usually about equal to the mean diameter, and can generally be used for it, avoiding the troublesome measurement of the short diameters.

The sizes of the separations of the sieves are always determined from the very last sand which passes through in the course of an analysis, and the results so obtained are quite accurate. With the elutriations average samples are inspected, and estimates made of the range in size of particles in each portion. Some stray particles both above and below the normal sizes are usually present, and even with the greatest care the result is only an approximation to the truth; still, a series of results made in strictly the same way should be thoroughly satisfactory, notwithstanding possible moderate errors in the absolute sizes.

Calculation of Results.—When a material has been separated into portions, each of which is accurately weighed, and the range in the sizes of grains in each portion determined, the weight of the particles finer than each size of separation can be calculated, and with enough properly selected separations the results can be plotted in the form of a diagram, and measurements of the curve taken for intermediate points with a fair degree of accuracy. This curve of results may be drawn upon a uniform scale, using the actual figures of sizes and of per cents by weight, or the logarithms of the figures may be used in one or both directions. The method of plotting is not of vital importance, and the method for any set of materials which gives the most easily and accurately drawn curves is to be preferred. In the diagram published in the Report of the Mass. State Board of Health for 1891, page 430, the logarithmic scale was used in one direction, but in many instances the logarithmic scale can be used to advantage in both directions. With this method it has been found that the curve is often almost a straight line through the lower and most important section, and very accurate results are obtained even with a smaller number of separations.

Examples of Calculation of Results.—Following are examples of representative analyses, showing the method of calculation used with the different methods of separation employed with various materials.

I. ANALYSIS OF A GRAVEL BY HAND-PICKING, 11,870 GRAMS TAKEN FOR ANALYSIS.

Number of Stones in Portion. (Largest Selected Stones.)	Total Weight of Portion. Grams.	Average Weight of Stones. Milligrams.	Estimated Weight of Smallest Stones. Milligrams.	Corresponding Size. Millimeters.	Total Weight of Stones Smaller than this Size.	Per Cent of Total Weight Smaller than this Size.
....	....	....	....	......	11,870	100
10..........	3,320	332,000	250,000	56	8,550	72
10..........	1,930	193,000	165,000	49	6,620	56
10..........	1,380	138,000	124,000	45	5,240	44
20..........	2,200	110,000	93,000	41	3,040	26
20..........	1,520	76,000	64,000	36	1,520	13
20..........	1,000	50,000	36,000	30	520	4.4
20..........	460	23,000	10,000	20	60	.5
10..........	40	4,000	2,000	11	20	.2
Dust........	20	....	....	......	......	......

The weight of the smallest stones in a portion given in the fourth column is estimated in general as about half-way between the average weight of all the stones in that portion and the average weight of the stones in the next finer portion.

The final results are **shown** by the figures in full-faced type in the last and **third** from the last columns. By plotting these figures we find that 10 per cent of the stones are less than 35 millimeters in diameter, and 60 per **cent are less than** 51 millimeters. The "uniformity coefficient," **as described below, is** the ratios of these numbers, or **1.46, while the** "effective size" **is 35** millimeters.

II. ANALYSIS OF A SAND BY MEANS OF SIEVES.

A portion of the sample was dried in a porcelain dish in an air-bath. Weight dry, 110.9 grams. It was put into a series of sieves in a mechanical shaker, and given one hundred turns (equal to about seven hundred single shakes). The sieves were then taken apart, and the portion passing the finest sieve weighed. After noting the weight, the sand remaining on the finest sieve, but passing all the coarser sieves, was added to the first and again weighed, this process being repeated until all the sample was upon the scale, weighing 110.7 grams, showing a loss by handling of only 0.2 gram. The figures were as follows:

Sieve Marked.	Size of Separation of this Sieve. Millimeters.	Quantity of Sand Passing. Grams.	Per Cent of Total Weight.	Sieve Marked.	Size of Separation of this Sieve. Millimeters.	Quantity of Sand Passing. Grams.	Per Cent of Total Weight.
190	.105	.5	.5	40	.46	56.7	51.2
140	.135	1.3	1.2	20	.93	89.1	80.5
100	.182	4.1	3.7	10	2.04	104.6	94.3
60	.320	23.2	21.0	6	3.90	110.7	100.0

Plotting the figures in heavy-faced type, we find from the curve that 10 and 60 per cent respectively are finer than .25 and .62 millimeter, and we have for effective size, as described above, .25, and for uniformity coefficient 2.5.

III. ANALYSIS OF A FINE MATERIAL WITH ELUTRIATION.

The entire sample, 74 grams, was taken for analysis. The sieves used were not the same as those in the previous analysis, and instead of mixing the various portions on the scale they were separately weighed. The siftings were as follows:

```
Remaining on sieve marked  10, above 2.2  millimeters....  1.5 grams
    "       "    "    "     20,   "   .98     "        ....  7.0   "
    "       "    "    "     40,   "   .46     "        .... 23.0   "
    "       "    "    "     70,   "   .24     "        .... 20.2   "
    "       "    "    "    140,   "   .13     "        ....  9.2   "
Passing sieve                140, below .13   "        .... 14.1   "
```

The 14.1 grams passing the 140 sieve were thoroughly mixed, and one third, 4.7 grams, taken for analysis. After ignition just below a red heat in a radiator, the weight was diminished by 0.47 gram. The portion above .08 millimeter and between .04 and .08 millimeter, separated as described above, weighed respectively 1.27 and 1.71 grams, and the portion below .04 millimeter was estimated by difference [4.7 − (0.47 + 1.27 + 1.71)] to be 1.25 grams. Multiplying these quantities by 3, we obtain the corresponding quantities for the entire sample, and the calculation of quantities finer than the various sizes can be made as follows:

Size of Grain.	Weight. Grams.	Size of Largest Particles. Millimeters.	Weight of all the Finer Particles. Grams.	Per Cent by Weight of all Finer Particles.
Above 2.20 millimeters........	1.50	....	74.00	100
.98–2.20 "	7.00	2.20	72.50	98
.46– .98 "	22.00	.98	65.50	89
.24– .46 "	20.20	.46	43.50	60
.13– .24 "	9.20	.24	23.30	32
.08– .13 "	3.81	.13	14.10	19
.04– .08 "	5.13	.08	10.29	14
.01– .04 "	3.75	.04	5.16	7
Loss on ignition (assumed to be less than .01 millimeter).....	1.41	.01	1.41	1.9

By plotting the heavy-faced figures we find that 10 and 60 per cent are respectively finer than .055 and .46 millimeter, and we have effective size .055 millimeter and uniformity coefficient 8.

The effective size and uniformity coefficient calculated in this way have proved to be most useful in various calculations, particularly in estimating the friction between the sands and gravels and water. The remainder of the article in the Report of the Mass. State Board of Health is devoted to a discussion of these relations which were mentioned in Chapter III of this volume.

APPENDIX IV.

A LIST OF SOME FILTRATION WORKS.

(The numbers in parenthesis indicate the years to which the statistics refer. (2) = 1892, (4) = 1894, etc.)

Place.	Population Supplied.	Filters. No.	Filters. Average Area. Acres.	Filters. Total Area. Acres.	Nominal Capacity of Works.	Average Daily Consumption for One Year.
Altona............	156,000	12	0.21	2.52 (4)	....	4,200,000 (2)
Amsterdam: Vecht	515,000 (4)	4	1.34	5.35 (4)	10,600,000	....
Dune.		7	0.68	4.80 (4)	....	6,300,000 (3)
Antwerp...........	200,000	8	0.26	2.10 (4)	....	1,720,000 (3)
Berlin: Müggel.....		22	0.57	12.70 (4)	23,000,000	....
Stralau.....	1,606,000	11	0.84	9.15 (4)	....	9,700,000 (3)
Tegel......		21	0.59	12.40 (4)	23,000,000	19,400,000 (3)
Bradford...........	364,000	6	0.77	4.62 (4)	....	12,000,000
Bremen............	146,000	...	...	2.35 (2)	....	2,850,000 (2)
Breslau............	335,000	5	1.02	5.12 (4)	....	7,300,000 (2)
Brunswick.........	105,000	...	...	1.46 (2)	....	2,070,000 (2)
Budapest..........	500,000	8	0.37	3.00 (4)	....	20,400,000 (2)
Choisy le Roi......	(100,000)	15	0.15	2.31 (4)	7,900,000	......
Copenhagen........	....	9	0.32	2.85 (4)	....	....
Darlington.........	40,000	7	0.19	1.32 (4)	....	2,000,000 (0)
Dordrecht.........	34,000	2	0.28	0.56 (4)	....	880,000 (3)
Dublin.	340,000	10	0.50	5.00 (4)	....	17,900,000
Edinburgh.........	794,000	4	0.50	2.00	....	....
Hague.............	170,000	6	0.48	2.88 (4)	....	4,200,000 (3)
Hamburg...........	583,000	18	1.89	34.00 (4)	....	31,600,000 (3)
Königsberg........	167,000	8	0.39	3.13 (4)	....	2,570,000 (2)
Liegnitz............	48,000	6	0.16	0.96 (4)	....	1,340,000 (3)
Liverpool: Oswestry Rivington	815,000	3	1.65	4.95 (4)	....	....
London, all filters...		113	1.02	115.75(4)	....	195,000,000*(3)
Chelsea..		7	0.96	6.75 (4)	....	13,400,000 (3)
E. London......		31	0.94	29.75 (4)	....	52,700,000 (3)
Grand Junction..		15	1.18	17.75 (4)	....	21,800,000 (3)
Lambeth.........	5,030,000	10	0.95	9.50 (4)	....	25,400,000 (3)
New River.......		20	0.82	16.50 (4)	....	44,800,000 (3)
Southw. & Vauxh.		18	1.14	20.50 (4)	....	36,600,000 (3)
W. Middlesex....		12	1.25	15.00 (4)	....	22,400,000 (3)
Lübeck............	64,000	7	0.36	2.54 (4)	....	3,400,000 (1)
Magdeburg.........	200,000	11	0.36	3.90 (4)	....	5,100,000 (2)
Middlesborough....	180,000	11	0.30	3.27 (4)	....	11,000,000 (0)
Rotterdam.........	240,000	18	0.35	6.30 (4)	....	13,300,000 (3)

* This is filtered water only; some of the companies secure some ground-water, which they mix with the filtered water, and which is included in the quantities for the separate companies, and as I have no records of these amounts to make the corrections, the latter quantities are slightly too great.

SOME FILTRATION WORKS—*Continued.*

Place.	Population Supplied.	Filters. No.	Filters. Average Area. Acres.	Filters. Total Area. Acres.	Nominal Capacity of Works.	Average Daily Consumption for One Year.
Schiedam	25,000	5	0.26	1.33 (4)	....	680,000 (3)
Stuttgart	139,000	7	0.15	1.05 (2)	....	2,500,000 (2)
St. Petersburg	960,000	11	0.53	5.85 (0)	....	39,000,000 (0)
Warsaw	500,000	12	0.52	6.20 (0)	....	6,100,000 (0)
York	70,000+	6	0.34	2.04 (4)	....	....
Zürich	100,000	7	0.17	1.18 (4)	....	5,500,000 (2)

STATISTICS OF OPERATION OF SOME FILTERS FOR ONE YEAR.

Place.	Year.*	Total Quantity, Millions of Gallons.	Area of Filters. Acres.	Average Daily Yield, Million Gallons per Acre.	Period between Scraping Days. Longest.	Period between Scraping Days. Shortest.	Period between Scraping Days. Mean.	Area of Filter-surface Cleaned. Acres.	Million Gallons per Acre Filtered between Scrapings.
Berlin, all filters	1892	9,600	21.55	1.23	....	....	....	....	....
" Tegel	1893	7,070	12.40	1.57	} 90	6	....	....	....
" Stralau	1893	3,510	9.15	1.05			....	....	....
Breslau	1892	2,660	4.12	1.76	74.2	19.0	37.3	38.1	70
Budapest	1892	7,360	3.00	6.70	10.33	0.79	3.3	254.0	29
Magdeburg	1892	1,835	2.48	2.02	55	8	25.1	32.3	57
Bremen	1892	1,040	2.35	1.22	35	7	23.4	37.0	28
Altona	1892	1,520	2.18	1.92	50	10	26.3	30.3	51
Königsberg	1892	1,000	1.98	1.39	47	7	20.0	28.8	35
Brunswick	1891	752	1.46	1.41	73	20	40	12.8	59
Stuttgart	1892	907	1.05	2.38	46	10	28	12.8	71
Stettin	1892	1,360	.95	3.90	17	4	9.5	31.3	44
Zürich	1891	1,995	.84	6.50	75	17	32	7.7	260
Posen	1892	340	.70	1.34	37	8	15.4	8.9	38
Lübeck	1891	1,225	.52	6.50	....	....	....	19.0	65
Frankfort	1891	254	.37	1.88	87	10	23.5	5.9	43
London, all filters	1892	65,783†	109.75	1.64	....	....	....	....	....
Chelsea	1892	4,164	6.75	1.69	....	....	....	....	157
E. London	1892	17,782	29.75	1.64	....	....	....	....	76
Grand Junction.	1892	8,165	17.75	1.26	....	....	....	....	73
Lambeth	1892	8,727	9.50	2.52	....	....	....	....	98
New River	1892	15,224	16.50	2.52	....	....	....	....	73
South. & Vaux.	1892	11,510	14.50	2.17	....	....	....	....	83
W. Middlesex	1892	7,660	15.00	1.40	....	....	....	....	71

*These results are for 12 months in every case, but not always for the calendar year. Many German results are for the year ending March 31.

† Filtered water only; see foot note on page 159.

APPENDIX V.

LONDON'S WATER-SUPPLY.

LONDON alone among great capitals is supplied with water by private companies. They are, however, under government supervision, and the rates charged for water are regulated by law. There are eight companies, each of which supplies its own separate district, so that there is no competition whatever. One of the companies supplying 460,000 people uses only ground-water drawn from deep wells in the chalk, but the other seven companies depend mainly upon the rivers Thames and Lea for their water. All water so drawn is filtered, and must be satisfactory to the water examiner, who is required to inspect the water supplied by each company at frequent intervals, and the results of the examinations are published each month.

In 1893 the average daily supply was 235,000,000 gallons, of which about 40,000,000 were drawn from the chalk, 125,000,000 from the Thames, and 70,000,000 from the Lea. Formerly some of the water companies drew water from the Thames within the city where it was grossly polluted, and the plagues and cholera which formerly ravaged London were in part due to this fact. These intakes were abandoned many years ago, and all the companies now draw their water from points outside of the city and its immediate suburbs.

The area of the watershed of the Thames above the intakes of the water companies is 3548 square miles, and the population living upon it in 1891 was 1,056,415. The Thames Conservancy Board has control of the main river for its whole length, and of all tributaries within ten miles in a straight line of the main river, but has no jurisdiction over the more remote feeders. The area drained is essentially agricultural, with but little manufacturing, and there are but few large towns. In the area coming under the conservators there are but six towns with populations above 10,000 and an aggregate population of 170,000, and there are but two or three other large towns on the remaining area more than ten miles from the river. These principal towns are as follows:

Town.	Population 1891.	Distance above Water Intakes.
Reading	60,054	49 miles
Oxford	45,791	87 "
New Swindon	27,295	116 "
High Wycomb	13,435	33 "
Windsor	12,327	18 "
Maidenhead	10,607	25 "
Guildford	14,319	20 "

Guildford is outside of the conservators' area. All of the above towns treat their sewage by irrigation.

Among the places that are regarded as the most dangerous are Chertsey and Staines, with populations of 9215 and 5060, only 8 and 11 miles above the intakes respectively. These towns are only partially sewered and still depend mainly on cesspools. An attempt is made to treat the little sewage which they produce upon land, but the work has not as yet been systematically carried out. There are also several small towns of 3000 inhabitants or less upon the upper river which do not treat their sewage so far as they have any, but, owing to their great distance, the danger from them is much less than from Chertsey and Staines. Twenty-one of the principal towns upon the watershed have sewage farms, and there are no chemical precipitation plants now in use.

Boats upon the river are not allowed to drain into it, but are compelled to provide receptacles for their sewage, and facilities are provided for removing and disposing of it; and as an additional precaution no boat is allowed to anchor within five miles of the intakes.

The conservators of the river Lea have control of its entire drainage area, which is about 460 square miles, measured from the East London water intakes, and has a population of 189,287. On this watershed there is but a single town with more than 10,000 inhabitants, this being Lutton near the headwaters of the river, with a population of 30,005. The sewage from Lutton and from seventeen smaller places is treated upon land. No crude sewage is known to be ordinarily discharged into the river. At Hereford, eleven miles above the East London intakes, there is a chemical precipitation plant. The conservators do not regard this treatment as satisfactory, and have recently conducted an expensive lawsuit against the local authorities to compel them to further treat their effluent. The suit was lost, the court holding that no actual injury to health had been shown. It is especially interesting to note that of the thirty-nine places on the Thames and the Lea giving their sewage systematic treatment there is but a single place using chemical precipitation, and there it is not considered satisfactory. Formerly quite a number of these towns used other processes than land treatment, but in every case but Hereford land treatment has been substituted.

In regard to the efficiency of the sewage farms, it is believed that in ordinary weather the whole of the sewage percolates through the land, and the inspectors of the Conservancy Boards strongly object to its being allowed to pass over the surface into the streams. The land, however, is for the most part impervious, as compared to Massachusetts and German sewage farms, and in times of heavy storms the land often has all the water it can take without receiving even the ordinary flow of sewage, and much less the increased storm-flow. At such times the sewage either does go over the surface, or perhaps more frequently is discharged directly into the rivers without even a pretence of treatment. The conservators apparently regard this as an unavoidable evil and do not vigorously oppose it. It is the theory that, owing to the increased dilution with the storm-flows, the matter is comparatively harmless, although it would seem that the reduced time required for it to reach the water-works intakes might largely offset the effect of increased dilution.

The water companies have large storage and sedimentation basins with an aggregate capacity equal to nine days' supply, but the proportion varies widely with the different companies. It is desired that the water held in reserve shall be alone used while the river is in flood, as, owing to its increased pollution, it is regarded as far more dangerous than the water at other times; but as no record is kept of the times when raw sewage is discharged, and no exact information is available in regard to the times when the companies do not take in raw water, it can safely be assumed that a considerable amount of raw sewage does become mixed with the water which is drawn by the companies.

The water drawn from the river is filtered through 113 filters having an area of 116 acres. None of the filters are covered, and with an average January temperature of 39° but little trouble with ice is experienced. A few new filters are provided with appliances for regulating the rate on each filter separately and securing regular and determined rates of filtration, but nearly all of the filters are of the simple type described on page 48, and the rates of filtration are subject to more or less violent fluctuation, the extent of which cannot be determined.

The area of filters is being continually increased to meet increasing consumption; the approximate areas of filters in use having been as follows:

1839	First filters built
1855	37 acres
1866	47 "
1876	77 "
1886	104 "
1894	116 "

There has been a tendency to reduce somewhat the rate of filtration. In 1868, with 51 acres of filters, the average daily quantity of water filtered was 111,000,000 gallons, or 2,180,000 gallons per acre. In 1884, with 97 acres of filter surface, the daily quantity filtered was 157,000,000 gallons, or 1,620,000 gallons per acre ; and in 1893, with 116 acres of filter surface and 195,000,000 gallons daily, the yield per acre was 1,680,000 gallons.

Owing to the area of filter surface out of use while being cleaned, the variations in consumption of water, and the imperfections of the regulating apparatus, the actual rates of filtration are often very much higher and at times may easily be double the figures given.

Evidence regarding the healthfulness of the filtered river-water was collected and examined in a most exhaustive manner in 1893 by a Royal Commission appointed to consider the water-supply of the metropolis in all its aspects with reference to future needs. This commission was unable to obtain any evidence whatever that the water as then supplied was unhealthy or likely to become so, and they report that the rivers can safely be depended upon for many years to come.

The numbers of deaths from all causes and from typhoid fever annually per million of inhabitants for the years 1885–1891 in the populations receiving their waters from different sources in London were as follows:

Water used.	Deaths from All Causes.	Deaths from Typhoid Fever.
Filtered Thames water only	19,501	125
" Lea water only	21,334	167
Kent wells only	18,001	123
Thames and Lea jointly	18,945	138
" " Kent jointly	18,577	133

The population supplied exclusively from the Lea by the East London Company is of a poorer class than that of the rest of London, and this may account for the slightly higher death-rate in this section. Aside from this the rate is remarkably uniform and shows no great difference between the section drinking ground-water only and those drinking filtered river-waters. The death-rate from typhoid fever is also very uniform and, although higher than that of some Continental cities with excellent water-supplies (Berlin, Vienna, Munich, Dresden), is very low—lower than in any American city of which I have records.

In this connection, it was shown by the Registrar-General that there is only a very small amount of typhoid fever on the watersheds of the Thames and Lea, so that the danger of infection of the water as distinct from pollution is less than would otherwise be the case. Thus for the seven years above mentioned the numbers of deaths from typhoid fever per million of population were only 105 and 120 on the watersheds of the Thames and the Lea respectively, as against 176 for the whole of England and Wales.

APPENDIX VI.

THE BERLIN WATER-WORKS.

THE original works were built by an English company in 1856, and were sold to the city in 1873 for $7,200,000.

The water was taken from the river Spree at the Stralau Gate, which was then above, but is now surrounded by, the growing city. The water was always filtered, and the original filters remained in use until 1893, when they were supplanted by the new works at Lake Müggel. Soon after acquiring the works the city introduced water from wells by Lake Tegel as a supplementary supply, but much trouble was experienced from crenothrix, an organism growing in ground-waters containing iron, and in 1883 this supply was replaced by filtered water from Lake Tegel. With rapidly-increasing pollution of the Spree at Stralau the purity of this source was questioned, and in 1893 it was abandoned (although still held as a reserve in case of urgent necessity), the supply now being taken from the river ten miles higher up, at Müggel.

The watershed of the Spree above Stralau, as I found by map measurement, is about 3800 square miles; the average rainfall is about 25 inches yearly. At extreme low water the river discharges 457 cubic feet per second, or 295 million gallons daily, and when in flood 5700 cubic feet per second may be discharged. The city is allowed by law to take 46 million gallons daily for water-supply, and this quantity can be drawn either at Stralau or at Müggel.

Above Stralau the river is polluted by numerous manufactories and washing establishments, and by the effluent from a considerable part of the city's extensive sewage farms. The shipping on this part of the river also is heavy, and sewage from the boats is discharged directly into the river. The average number of bacteria in the Spree at this point is something over ten thousand per cubic centimeter, and 99.6 per cent of them were removed by the filters in 1893.

The watershed of the Spree above the new water-works at Müggel I found by map measurement to be 2800 square miles, and the low water-discharge is said to be 269 million gallons daily. The river at this point flows through Lake Müggel, which forms a natural sedimentation-basin, and the raw water is quite clear except in windy weather.

There were 16 towns on the watershed with populations above 2000 each in 1890, and an aggregate population of 132,000, which does not include the population of the smaller places or country districts. None of these places purify their sewage so far as they have any. Fürstenwalde with a population of 12,935, and 22 miles above Müggel, has surface sewers discharging directly into the river. Above Fürstenwalde the river runs through numerous lakes which probably remove the effect of the pollution from the more distant cities. There is considerable shipping on the river for some miles above Fürstenwalde (which forms a section of the Friedrich Wilhelm Canal), but hardly any between Müggel and Fürstenwalde. The raw water at Müggel contains two or three hundred bacteria per cubic centimeter, and is thus a comparatively pure water before filtration. It is slightly peaty and the filtered water has a light straw color.

Lake Tegel, which supplies the other part of the city's supply, is an enlargement of the river Havel. The watershed above Tegel I find to be about 1350 square miles, and the annual rainfall is about 22 inches. The low water-discharge is said to be 182 million gallons daily, and the city is allowed by law to take 23 million gallons for water-supply.

There were ten towns upon the watershed with populations above 2000 each in 1890, and with an aggregate population of 44,000. Of these Tegel is directly upon the lake with a population of 3000, and Oranienburg, 14 miles above, has a population of 6000 and is rapidly increasing. The shipping on the lake and river is heavy. The lake water ordinarily contains two or three hundred bacteria per cubic centimeter. The lake is shallow and becomes turbid in windy weather.

There are 21 filter-beds at Tegel with a combined area of 12.40 acres to furnish a maximum of 23 million gallons of water daily, and 22 filters at Müggel with a combined area of 12.7 acres to deliver the same quantity. Twenty-two more filters will be built at Müggel within a few years to purify the full quantity which can be taken from the river. All of these filters are covered with brick arches supported by pillars about 16 feet apart from centre to centre in each direction, and the whole is covered by nearly 3 feet of earth, making them quite frost-proof. The original filters at Stralau were open, but much difficulty was experienced with them in winter.

The bottom of the filters at Tegel consists of 8 inches of concrete above 20 inches of packed clay and with 2 inches of cement above, and slopes slightly from each side to the centre. The central drain goes the whole length of the filters and has a uniform cross-section of about $\frac{1}{7300}$ of the area of the whole bed. There are no lateral drains, but the water is brought to the central drain by a twelve-inch layer of stones as large as a man's fist; above this there is another foot of gravel of graded sizes supporting two feet of fine sand, which is reduced by scraping to half its thickness before the sand is replaced. The average depth of water above the sand is nearly 5 feet. The filters are not allowed to filter at a rate above 2.57 million gallons per acre daily, and at this rate with 70 per cent of the area in service the whole legal quantity of water can be filtered. The filters work at precisely the same rate day and night, and the filtered water is continuously pumped as filtered to ample storage reservoirs at Charlottenburg. The pumps which lift the water from the lake to the filters work against a head of 14 feet. The apparatus for regulating the rate of filtration was described on page 51.

As yet no full description of the Müggel works has been published, but they resemble closely the Tegel works. Both were designed by or under the direction of the late director of the water-works, Mr. Henry Gill.

The average daily quantity of water supplied for the fiscal year ending March 31, 1893, was 29,000,000 gallons daily, which estimate allows 10 per cent for the slip of the pumps. Of this quantity 9,650,000 was furnished by Stralau and 19,350,000 by Tegel. The greatest consumption in a single day was 43,300,000 gallons, or 26.6 gallons per head, while the average quantity for the year was 18.4 gallons per head. All water without exception is sold by meter, the prices ranging from 27.2 cents a thousand gallons for small consumers to 13.6 cents for large consumers and manufacturers. The average receipts for all water pumped, including that used for public purposes and not paid for, were 15.4 cents a thousand gallons, against the cost of production, 9.8 cents, which covers operating expenses, interest on capital, and provision for sinking fund. This leaves a handsome net profit to the city. On account of the comparatively high price of the city water and the ease with which well-water is obtained, the latter is almost exclusively used for running engines, manufacturing purposes, etc., and this in part explains the very low per-capita consumption.

The volume of sewage, however, for the same year, including rain-water, except during heavy showers, was only 29 gallons per head, showing even with the private water-supplies an extraordinarily low consumption.

The friction of the water in the 4.75 miles of 3-foot pipe between Tegel and the reservoir at Charlottenburg presents an interesting point. When well-water with crenothrix was pumped, the friction rose to 34.5 feet, when the velocity was 2.46 feet per second. According to Herr Anklamm, who had charge of the works at the time, the friction was reduced to 19.7 feet when filtered water was used and after the pipe had been flushed, and this has not increased with continued use. He calculated the friction for the velocity according to Darcy 15.0 feet, Lampe 17.8 feet, Weisbach 18.7 feet, and Prony 21.5 feet.

APPENDIX VII.

ALTONA WATER-WORKS.

THE Altona water-works are specially interesting as an example of a water drawn from a source polluted to a most unusual extent: the sewage from cities with a population of 770,000, including its own, is discharged into the river Elbe within ten miles above the intake and upon the same side.

The area of the watershed of the Elbe above Altona is about 52,000 square miles, and the average rainfall is estimated to be about 28 inches, varying from 24 or less near its mouth to much higher quantities in the mountains far to the south. On this watershed there are 46 cities, which in 1890 had populations of over 20,000 each, and in addition there is a permanent population upon the river-boats estimated at 20,000, making in all 5,894,000 inhabitants, without including either country districts or the numberless cities with less than 20,000 inhabitants each. The sewage from about 1,700,000 of these people is purified before being discharged; and assuming that as many people living in cities smaller than 20,000 are connected with sewers as live in larger places without being so connected, the sewage of over four million people is discharged untreated into the Elbe and its tributaries.

The more important of these sources of pollution are the following:

City.	Population in 1890.	On what River.	Approximate Distance, Miles
Shipping................	20,000		
Altona	143,353	Elbe	6
Hamburg	570,534	"	7
Wandsbeck.............	20,586	"	8

City.	Population in 1890.	On what River.	Approximate Distance, Miles.
Harburg	35,101	Elbe	11
Magdeburg	202,325	"	185
Dresden	276,085	"	354
Berlin and suburbs	1,787,859	Havel	243
Halle	101,401	Saale	272
Leipzig	355,485	Elster	305
Chemnitz	138,955	Mulde	340
Prague	310,483	Moldau	500

The sewage of Berlin and of most of its suburbs is treated before being discharged, and in addition the Havel flows through a series of lakes below the city, allowing better opportunities for natural purification than in the case of any of the other cities. Halle treats less than a tenth of its sewage. Magdeburg will treat its sewage in the course of a few years. Leipzig, Chemnitz, and other places are thinking more or less seriously of purification.

The number of bacteria in the raw water at Altona fluctuates with the tide and is extremely variable; numbers of 50,000 and 100,000 are not infrequent, but 10,000 to 40,000 is perhaps about the usual range.

The works were originally built by an English company in 1860, and have since been greatly extended. They were bought by the city some years ago. The water is pumped directly from the river to a settling-basin upon a hill 280 feet above the river. From this it flows by gravity through the filters to the slightly lower pure-water reservoir and to the city without further pumping. The filters are open, with nearly vertical masonry walls, as described in Kirkwood's report. The cross-section of the main underdrain is $\frac{1}{2800}$ of the area of the beds.

Considerable trouble has been experienced from frost. With continued cold weather it is extremely difficult to satisfactorily scrape the filters, and very irregular rates of filtration may result at such times. In the last few years, with systematic bacterial investigation, it has been found that greatly decreased efficiency frequently follows continued cold weather, and the mild epidemics of typhoid fever from which the city has long suffered have generally occurred after these times. Thus a light epidemic of typhoid in 1886 came in March, following a light epidemic in Hamburg. In 1887 a severe epidemic in February followed a severe epidemic in Hamburg in December and January. In 1888 a severe epidemic in March followed an epidemic in Hamburg lasting from November to January. Hamburg's epidemic of 1889, coming in warm weather, September and October, was followed by only a very slight increase in Altona. In 1891 Altona suffered again in February from a severe epidemic, although very little typhoid had been in Hamburg. A less severe outreak also came in February, 1892, and a still slighter one in February, 1893. In the ten years 1882–1892, of five well-marked epidemics, three broke out in February and two in March, while two smaller outbreaks came in December and January. No important outbreak has ever occurred in summer or in the fall months, when typhoid is usually most prevalent, thus showing clearly the bad effect of frost upon open filters (see Appendix II). With steadily increasing consumption the sedimentation-basin capacity of late years has become insufficient as well as the filtering area, and it is not unlikely that with better conditions a much better result could be obtained in winter even with open filters.*

The brilliant achievement of the Altona filters was in the summer of 1892, when they protected the city from the cholera which

---

* In the *Centralblatt für Bakteriologie*, 1895, page 881, Reinsch discusses at length the cause of the inferior results at Altona in winter, and has apparently discovered a new factor in producing them. Owing to defective construction of the outlets for the sedimentation-basins they have failed to act properly in presence of excessive quantities of ice, and the sediment from the basins has been discharged in large quantity upon the filters, and a small fraction of the many millions of bacteria in it have passed through the filters. He has experimented with this sediment applied to small filters, and has become convinced that to secure good work under all conditions a much deeper layer of sand than that generally considered necessary must be used, and his work emphasizes the importance of the action of the sand in distinction from the action of the sediment layer, which has often been thought to be the sole, or at least the principal, requirement of good filtration.

so ravaged Hamburg, although **the raw water at Altona must have** contained **a vastly** greater quantity of infectious matter **than that** which worked such **havoc** in Hamburg.

From these records it appears that for about nine months of the **year the** Altona filters protect the city from the impurities of the Elbe **water, but** that during **cold weather,** with continued mean **temperatures** below the freezing-point, such protection is not completely **afforded, and bad** effects have occasionally resulted. Notwithstanding the recent construction of open filters in Hamburg it appears to **me** that there must always be **more or less** danger **from open filters in such a** climate. Hamburg's danger, however, will **be much** less than Altona's on account of its better intake above the **outlets** of the sewers of Hamburg and Altona, which are the most important points of pollution **at Altona.**

APPENDIX VIII.

HAMBURG WATER-WORKS.

THE source and quality of the water previously supplied has been sufficiently indicated in Appendix II. It was originally intended to filter the water, but the construction of filters was postponed from time to time until the fall of 1890, when the project was seriously taken up, and work was commenced in the spring of 1891. Three years were allowed for construction. In 1892, however, the epidemic of cholera came, killing 8605 residents and doing incalculable damage to the business interests of the city. The health authorities found that the principal cause of this epidemic was the polluted water-supply. To prevent a possible recurrence of cholera in 1893, the work of construction of the filters was pressed forward much more rapidly than had been intended. Electric lights were provided to allow the work to proceed nights as well as days, and as a result the plant was put in operation May 27, 1893, a full year before the intended time. Owing to the forced construction the cost was materially increased.

The new works take the raw water from a point one and a half miles farther up-stream, where it is believed the tide can never carry the city's own sewage, as it did frequently to the old intake. The water is pumped from the river to settling-basins against heads varying with tide and the water-level in the basins from 8 to 22 feet. Each of the four settling-basins has an area of about 10 acres, and, with the water 6.56 feet deep, holds 20,500,000 gallons, or 82,000,000 gallons in all. The works are intended to supply a maximum of 48,000,000 gallons daily, but the present average consumption is only about 35,000,000 gallons (1892), or 59 gallons per head for 600,000 population. This consumption is regarded as excessive, and it is hoped that it will be reduced materially by the more general use of meters. The sedimentation-basins are surrounded by earthen embankments with slopes of 1 : 3, the inner sides being paved with brick above a clay layer. The water flows by gravity from these basins to the filters, a distance of 1½ miles, through a conduit 8½ feet in diameter. The flow of the water out of the basins and from the lower end of the conduit is regulated by automatic gates connected with floats, shown by Fig. 11, page 56.

The filters are 18 in number, and each has an effective area of 1.89, or 34 acres in all. They are planned to filter at a rate of 1.60 million gallons per acre daily, which with 16 filters in use gives a daily quantity of 48,000,000 gallons as the present limit of the works. The sides of the filters are embankments with 1 : 2 slopes. Both sides and bottoms have 20 inches of packed clay, above which are 4 inches of puddle, supporting a brick pavement laid in cement. The bricks are laid flat on the bottom, but edgewise on the sides where they will come in contact with ice.

The main effluent-drain has a cross-section for the whole length of the filter of 4.73 square feet, or $\frac{1}{17300}$ of the area of the filter; and even at the low rate of filtration proposed, the velocity in the drain will reach 0.97 foot. The drain has brick sides, 1.80 feet high, covered with granite slabs. The lateral drains are all of brick with numerous large openings for admission of water. They are not ventilated, and I am unable to learn that any bad results follow this omission.

The filling of the filters consists of 2 feet of gravel, the top being of course finer than the bottom layers, above which are 40 inches of sand, which are to be reduced to 24 inches by scraping before being refilled. The water over the sand, when the latter is of full depth, is 43 inches deep, and will be increased to 59 inches with the minimum sand-thickness. The apparatus for regulating the rate of filtration was described page 52. The cost of the entire plant, including 34 acres effective filter-surface, 40 acres of sedimentation-basins, over 2 miles of 8½-foot conduit, pumping-machinery, sand-washing apparatus, laboratory, etc., was about 9,500,000 marks, or $2,280,000. This all reckoned on the effective filter area is $67,000 per acre, or $3.80 per head for a population of 600,000.

The death-rate since the introduction of filtered water has been lower than ever before in the history of the city, but as it is thought that other conditions may help to this result, no conclusions are as yet drawn.

APPENDIX IX.

NOTES ON SOME OTHER EUROPEAN WATER-SUPPLIES.

Amsterdam.—The water is derived from open canals in the dunes. These canals have an aggregate length of about 15 miles, and drain about 6200 acres. The water, as it enters the canals from the fine dune-sand, contains iron, but this is oxidized and deposited in the canals. The water after collection is filtered. It has been suggested that by using covered drains instead of open canals for collecting the water, the filtration would be unnecessary; but, on the other hand, the cost of building and maintaining covered drains in the very fine sand would be much greater than that of the canals, and it is believed, also, that the water so collected would contain iron, the removal of which might prove as expensive as the present filtration. In 1887 filters were built to take water from the river Vecht, but the city has refused to allow the English company which owns the water-works to sell this water for domestic purposes, and it is only used for public and manufacturing purposes, only a fraction of the available supply being required. Leyden, the Hague, and some other Dutch cities have supplies like the dune supply of Amsterdam, and they are invariably filtered.

Antwerp is also supplied by an English company. The raw water is drawn from a small tidal river, which at times is polluted by the sewage of Brussels. It is treated by metallic iron in Anderson revolver purifiers, and is afterward filtered at a rather low average rate. The hygienic results are closely watched by the city authorities, and are said to be satisfactory.

Rotterdam.—The raw water is drawn from the Maas, as the

Dutch call the main stream of the Rhine after it crosses their border. The population upon the river and its tributaries in Switzerland, Germany, Holland, France, and Belgium is very great; but the flow is also great, and the low water flow is exceptionally large in proportion to the average flow, on account of the melting snow in summer in Switzerland, where it has its origin.

The original filters had wooden under-drains, and there was constant trouble with crenothrix until the filters were reconstructed without wood, since which time there has been no farther trouble. The present filters are large and well managed. There is ample preliminary sedimentation.

Schiedam.—The filters at Schiedam are comparatively small, but are of unusual interest on account of the way in which they are operated. The intake is from the Maas just below Rotterdam. The city was unable to raise the money to seek a more distant source of supply, and the engineer, H. P. N. Halbertsma, was unwilling to recommend a supply from so doubtful a source without more thorough treatment than simple sand-filtration was then thought to be. The plan adopted is to filter the supply after preliminary sedimentation through two filters of 0.265 acre each, and the resulting effluent is then passed through three other filters of the same size. River sand is used for the first, and the very fine dune sand for the second filtration. The cost both of construction and operation was satisfactory to the city, and much below that of any other available source; and the hygienic results have been equally satisfactory, notwithstanding the unfavorable position of the intake.

Magdeburg.—The supply is drawn from the Elbe, and is filtered through vaulted filters after preliminary sedimentation. The pollution of the river is considerable, although less than at Altona or even at Hamburg. The city has been troubled at times by enormous discharges of salt solution from salt-works farther up, which at extreme low water have sometimes rendered the whole river brackish and unpleasant to the taste; but arrangements have now been made which, it is hoped, will prevent the recurrence of this trouble.

Breslau is supplied with filtered water from the river Oder, which has a watershed of 8200 square miles above the intake, and is polluted by the sewage from cities with an aggregate population of about 200,000, some of which are in Galicia, where cholera is often prevalent. In recent years the city has been free from cholera, and from more than a very limited number of typhoid-fever cases; but the pollution is so great as to cause some anxiety, notwithstanding the favorable record of the filters, and there is talk of the desirability of securing another supply. Until 1893 there were four filter-beds, with areas of 1.03 acres each, and not covered. In 1893 a fifth bed was added. This is covered by vaulting and is divided into four sections, which are separately operated, so that it is really four beds of 0.25 acre each. The vaulting is concrete arches, supported by steel I beams in one direction.

Budapest.—A great variety of temporary water-supplies have at different times been used by this rapidly growing city. The filters which for some years have supplied a portion of the supply have not been altogether satisfactory; but perhaps this was due to lack of preliminary sedimentation for the extremely turbid Danube water, and also to inadequate filter-area. The city is rapidly building and extending works for a supply of ground-water, and in 1894 the filters were only used as was necessary to supplement this supply, and it was hoped that enough well-water would be obtained to allow the filters to be abandoned in the near future. The Danube above the intake receives the sewage of Vienna and innumerable smaller cities, but the volume of the river is very great compared to other European streams, so that the relative pollution is not so great as in many other places.

Zürich.—The raw water is drawn by the city from the Lake of Zürich near its outlet, and but a few hundred feet from the heart of the city. Although no public sewers discharge into the lake, there is some pollution from boats and bathers and other sources, and, judging by the number of bacteria in the raw water, this pollution is increasing. The raw water is extremely free from sediment, and the filters only become clogged very slowly. The rate of filtration is high, habitually reaching 7,000,000 gallons per acre daily; but, with the clear lake water and long periods between scrapings, the results are excellent even at this rate. The filters are all covered with concrete groined arches.

Filtration was commenced in 1886, and was followed by a sharp decline in the amount of typhoid fever, which, up to that time, had been rather increasing; for the six years before the change there were sixty-nine deaths from this cause annually per 100,000 living, and for the six years after only ten, or one seventh as many; and this reduction is attributed by the local authorities to the filtration.*

St. Petersburg.—The supply is drawn from the Neva River by an English company, and is filtered through vaulted filters at a very high rate.

Warsaw.—The supply is drawn from the Weichsel River by the city, and is filtered through vaulted filters after preliminary sedimentation at a rate never exceeding 2,570,000 gallons per acre daily.

THE USE OF UNFILTERED SURFACE-WATERS.

The use of surface-water without filtration in Europe is comparatively limited. In Germany this use is now prohibited by the Imperial Board of Health. In Great Britain, Glasgow draws its supply unfiltered from Loch Katrine; and Manchester and some other towns use unfiltered waters from lakes or impounding reservoirs the watersheds of which are entirely free from population. The best English practice, however, as in Germany, requires the filtration of such waters even if they are not known to receive sewage, and the

---

* Licht- u. Wasserwerke, Zürich, 1892, page 32.

unpolluted supplies of Liverpool, Bradford, Dublin, and **many other cities are filtered before use.**

THE USE OF GROUND-WATER.*

Ground-waters are extensively used in Europe, and apparently in some localities the geological formations are unusually favorable to this kind of supply. Paris derives all the water it now uses for domestic purposes from springs, but has a supplementary supply from the river for other purposes. Vienna and Munich also obtain their entire supplies from springs, while Budapest, Cologne, Leipzig, Dresden, Frankfurt, many of the great French cities, Brussels, a part of London, and many other English cities derive their supplies from wells or filter-galleries, and among the smaller cities all over Europe ground-water supplies are more numerous than other kinds.

---

* Descriptions of some of the leading European ground-water supplies were given by the author in the Jour. **Asso.** Eng. Soc., Feb. 1895, p. **113.**

APPENDIX X.

LITERATURE OF FILTRATION.

THE following is a list of a number of articles on filtration. The list is not complete, but it is believed that it contains the greater part of articles upon slow sand-filtration, and that it will prove serviceable to those who wish to study the subject more in detail.

ANKLAMM. Glasers Annalen, 1886, p. 48.
    A description of the Tegel filters at Berlin, with excellent plans.

BAKER. Engineering News.
    Water purification in America: a series of descriptions of filters, as follows : Aug. 3, 1893, Lawrence filter and description of apparatus of screening sand and gravel; Apr. 26, 1894, filter at Nantucket, Mass.; June 7, 1894, filters at Ilion, N. Y., plans; June 14, 1894, filters at Hudson, N. Y.; July 12, 1894, filters at Zürich, Switzerland, plans ; Aug. 23, 1894, filters at Mt. Vernon, N. Y., plans.

BERTSCHINGER. Journal für Gas- und Wasserversorgung, 1889, p. 1126.
    A record of experiments made at Zürich upon the effect of rate of filtration, scraping, and the influence of vaulting. Rate and vaulting were found to be without effect, but poorer results followed scraping. The numbers of bacteria in the lake-water were too low to allow conclusive results.

——— Journal für Gas- und Wasserversorgung, 1891, p. 684.
    A farther account of the Zürich results, with full analyses and a criticism of Fränkel and Piefke's experiments.

BOLTON. Pamphlet, 1884.
    Descriptions and statistics of London filters.

BÖTTCHER and OHNESORGE. Zeitschrift für Bauwesen, 1876, p. 343.
    A description of the Bremen works, with full plans.

BURTON. Water-supply of Towns. London, 1894.
    Pages 94–115 are upon filtration and mention a novel method of regulating the rate.

CODD. Engineering News, Apr. 26, 1894.
    A description of a filter at Nantucket, Mass.

CRAMER. Centralblatt für Bauwesen, 1886, p. 42.
  A description of filters built at Brieg, Germany.
CROOK. London Water-supply. London, 1883.
DELBRUCK. Allgemeine Bauzeitung, 1853, p. 103.
  A general article on filtration ; particularly valuable for notices of early attempts at filtration and of the use of alum.
Deutsche Verein von Gas- und Wasserfachmänner.
  Stenographic reports of the proceedings of this society are printed regularly in the *Journal für Gas- und Wasserversorgung*, and the discussions of papers are often most interesting.
DROWN. Journal Association Eng. Societies, 1890, p. 356.
  Filtration of natural waters.
FISCHER. Vierteljahresschrift für Gesundheitspflege, 1891, p. 82.
  Discussion of papers on water-filtration.
FRÄNKEL. Vierteljahresschrift für Gesundheitspflege, 1891, p. 38.
  On filters for city water-works.
FRÄNKEL and PIEFKE. Zeitschrift für Hygiene, 1891, p. 38, Leistungen der Sandfiltern.
E. FRANKLAND. Report in regard to the London filters for 1893 in the Annual Summary of Births, Deaths, and Causes of Death in London and Other Great Towns, 1893. Published by authority of the Registrar-General.
P. FRANKLAND. Proc. Royal Society, 1885, p. 379.
  The removal of micro-organisms from water.
——— Proceedings Inst. Civil Engineers, 1886, lxxxv. p. 197.
  Water-purification ; its biological and chemical basis.
——— Trans. of Sanitary Institute of Great Britain, 1886.
  Filtration of water for town supply.
FRÜHLING. Handbuch der Ingenieurwissenschaften, vol. ii.
  Chapter on water-filtration gives general account of filtration, with details of Königsberg filters built by the author and not elsewhere published.
FULLER. Report Mass. State of Board of Health, 1892, p. 449.
    "   "   "   "   "   "   "   1893, p. 453.
  Accounts of the Lawrence experiments upon water-filtration for 1892 and 1893.
——— American Public Health Association, 1893, p. 152.
  On the removal of pathogenic bacteria from water by sand filtration.
——— American Public Health Association, 1894, p. 64.
  Sand filtration of water with special reference to results obtained at Lawrence, Mass.

GILL.   Deutsche Bauzeitung, 1881, p. 567.
    On American rapid filters. The author shows that they are not to be thought of for Berlin, as they would be more expensive as well as probably less efficient than the usual procedure.
——— Journal für Gas- und Wasserversorgung, 1892, p. 596.
    A general account of the extension of the Berlin filters at Müggel. No drawings.
GRAHN.   Journal für Gas- und Wasserversorgung, 1877, p. 543.
    On the filtration of river-waters.
——— Journal für Gas- und Wasserversorgung, 1890, p. 511.
    Filters for city water-works.
——— Vierteljahresschrift für Gesundsheitpflege, 1891, p. 76.
    Discussion of papers presented on filtration.
——— Journal für Gas- und Wasserversorgung, 1894, p. 185.
    A history of the " Rules for Water-filtration " (Appendix I), with some discussion of them.
GRAHN and MEYER.   Reiseberichte über künstliche central Sandfiltration. Hamburg, 1876.
    An account of the observations of the authors in numerous cities, especially in England.
GRENZMER.   Centralblatt der Bauverwaltung, 1888, p. 148.
    A description of new filters at Amsterdam, with plans.
GRUBER.   Centralblatt für Bacteriologie, 1893, p. 488.
    Salient points in judging of the work of sand-filters.
HALBERTSMA.   Journal für Gas- und Wasserversorgung, 1892, p. 43.
    Filter-works in Holland. Gives sand, gravel, and water thickness, with diagrams.
——— Journal für Gas- und Wasserversorgung, 1892, p. 686.
    Description of filters built by the author at Leeuwarden, Holland, with plans.
HART.   Proceedings Inst. of Civil Engineers, 1890, c. p. 217.
    Description of filters at Shanghai.
HAUSEN.   Journal für Gas- und Wasserversorgung, 1892, p. 332.
    An account of experiments made for one year with three 16-inch filters at Helsingfors, Finland, with weekly analyses of effluents.
HAZEN.   Report of Mass. State Board of Health, 1891, p. 601.
    Experiments upon the filtration of water.
——— Report of Mass. State Board of Health, 1892, p. 539.
    Physical properties of sands and gravels with reference to their use in filtration. (Appendix III.)
HUNTER.   Engineering, 1892, vol. 53, p. 621.
    Description of author's sand-washing apparatus.

KIRKWOOD. Filtration of River-water. New York, 1869.

A report upon European filters for the St. Louis Water Board in 1866. Contains a full account of thirteen filtration-works visited by the author, and of a number of filter-galleries, with a project for filters for St. Louis. This project was never executed, but the report is a wonderful work which appeared a generation before the American public was able to appreciate it. It was translated into German, and the German edition was widely circulated and known.

KOCH. Zeitschrift für **Hygiene,** 1893.

Water-filtration and Cholera: a discussion of the Hamburg epidemic of 1892 in reference to the effect of filtration.

KRÖHNKE. Journal für Gas- und Wasserversorgung, 1893, p. **513.**

An **account of** experiments made at Hamburg, as a result of which the author recommends the addition of cuprous chloride to the water before filtration to secure greater bacterial efficiency.

KÜMMEL. Journal für Gas- und Wasserversorgung, 1877, p. 452.

Operation of the Altona filters, with analyses.

——— Vierteljahresschrift für Gesundheitspflege, 1881, p. 92.

The water-works of the city of Altona.

——— Journal für Gas- und Wasserversorgung, 1887, p. 522.

An article opposing the use of rapid filters (David's process).

——— Journal für Gas- und Wasserversorgung, 1890, p. 531.

A criticism of Fränkel and Piefke's results, with some statistics of German and English filters. (The English results are taken without credit from Kirkwood.)

——— Vierteljahresschrift für Gesundheitspflege, 1891, p. 87.

Discussion of papers on filtration, with some statistics.

——— Vierteljahresschrift für Gesundheitspflege, 1892, p. 385.

The epidemic of typhoid-fever in Altona in 1891.

——— Journal **für Gas-** und Wasserversorgung, 1893, p. **161.**

Results of experiments upon filtration made **at Altona, and** bacterial **results** of **the** Altona filters in connection **with** typhoid death-rates.

——— Trans. Am. Society of Civil **Engineers, 1893, xxx. p.** 330.

Questions of water-filtration.

LESLIE. Trans. Inst. Civil Engineers, 1883, lxxiv. p. **110.**

A short description of filters at Edinburgh.

LINDLEY. A report for the commissioners of the Paris Exposition of 1889 upon the purification of river-waters, and published in French or **German in a** number of journals, **among** them *Journal für Gas- und Wasserversorgung,* 1890, p. **501.**

This is a most satisfactory discussion of the conditions which modern experience has shown to be essential to successful filtration.

MASON. Engineering News, Dec. 7, 1893.
Filters at Stuttgart, Germany, with plans.

MEYER and SAMUELSON. Deutsche Bauzeitung, 1881, p. 340.
Project for filters for Hamburg, with diagrams. Except in detail, this project is the same as that executed twelve years later.

MEYER. Deutsche Bauzeitung, 1892, p. 519.
Description of the proposed Hamburg filters, with diagrams.

——— The Water-works of Hamburg.
A paper presented to the International Health Congress at Rome, March 1894, and published as a monograph. It contains a full description of the filters as built, with drawings and views in greater detail than the preceding paper.

MILLS. Special Report Mass. State Board of Health on the Purification of Sewage and Water, 1890, p. 601.
An account of the Lawrence experiments, 1888-1890.

——— Report Mass. State Board of Health, 1893, p. 543.
The Filter of the Water-supply of the City of Lawrence and its Results.

——— Trans. Am. Society of Civil Engineers, 1893, xxx. p. 350.
Purification of Sewage and Water by Filtration.

NEVILLE. Engineering, 1878, xxvi. p. 324.
A description of the Dublin filters, with plans.

NICHOLS. Report Mass. State Board of Health, 1878, p. 137.
The filtration of potable water.

OESTER. Gesundheits-Ingenieur, 1893, p. 505.
What is the Rate of Filtration? A purely theoretical discussion.

ORANGE. Trans. Inst. Civil Engineers, 1890, c. p. 268.
Filters at Hong Kong.

PFEFFER. Deutsche Bauzeitung, 1880, p. 399.
A description of filters at Liegnitz, Germany.

PIEFKE. Results of Natural and Artificial Filtration. Berlin, 1881. Pamphlet.

——— Journal für Gas- und Wasserversorgung, 1887, p. 595. Die Principien der Reinwassergewinnung vermittelst Filtration.
A sketch of the theory and practical application of filtration.

——— Zeitschrift für Hygiene, 1889, p. 128. Aphorismen über Wasserversorgung.
A discussion of the theory of filtration, with a number of experiments on the thickness of sand-layers, etc.

PIEFKE. Vierteljahresschrift für Gesundheitspflege, 1891, p. 59.
On filters for city water-works.
FRÄNKEL and PIEFKE. Zeitschrift für Hygiene, 1891, p. 38.
Leistungen der Sandfiltern. An account of the partial obstruction of the Stralau filters by ice, and a typhoid epidemic which followed. Experiments were then made upon the passage of cholera and typhoid germs through small filters.
PIEFKE. Journal für Gas- und Wasserversorgung, 1891, p. 208. Neue Ermittelungen über Sandfiltration.
The above mentioned experiments being objected to on certain grounds, they were repeated by Piefke alone, confirming the previous observations on the passage of bacteria through filters, but under other conditions.
———— Zeitschrift für Hygiene, 1894, p. 151. Über Betriebsführung von Sandfiltern.
A full account of the operation of the Stralau filters in 1893, with discussion of the efficiency of filtration, etc.
PLAGGE AND PROSKAUER. Zeitschrift für Hygiene, II. p. 403.
Examination of water before and after filtration at Berlin, with theory of filtration.
REINCKE. Bericht über die Medicinische Statistik des Hamburgischen Staates für 1892.
Contains a most valuable discussion of the relations of filtration to cholera, typhoid fever, and diarrhœa, with numerous tables and charts. (Abstract in Appendix II.)
REINSCH. Centralblatt für Bakteriologie, 1895, p. 881.
An account of the operation of the Altona filters. High numbers of bacteria in the effluents have often resulted from the discharge of sludge from the sedimentation-basins onto the filters, due to the interference of ice on the action of the floating outlet for the basins, and this, rather than the direct effect of cold, is believed to be the direct cause of the low winter efficiency. The author urges the necessity of a deeper sand-layers in no case less than 18 inches thick.
RENK. Gesundheits-Ingenieur, 1886, p. 54.
———— Über die Ziele der künstliche Wasserfiltration.
RUHLMANN. Wochenblatt für Baukunde, 1887, p. 409.
A description of filters at Zürich.
SALBACH. Glaser's Annalen, 1882.
Filters at Groningen, Holland, built in 1880. Alum used.
SAMUELSON. Translation of Kirkwood's "Filtration of River-waters" into German, with additional notes especially on the theory of filtration and the sand to be employed. Hamburg, 1876.

SAMUELSON. Filtration and constant water-supply. Pamphlet. Hamburg, 1882.
——— Journal f. Gas- und Wasserversorgung, 1892, p. 660.
  A discussion of the best materials and arrangement for sand-filters.
SCHMETZEN. Deutsche Bauzeitung, 1878, p. 314.
  Notice and extended criticism of Samuelson's translation of Kirkwood.
SEDDEN. Jour. Asso. Eng. Soc., 1889, p. 477.
  In regard to the sedimentation of river-waters.
SEDGWICK. New England Water-works Association, 1892, p. 103.
  European methods of Filtration with Reference to American Needs.
SOKAL. Wochenschrift der östreichen Ingenieur-Verein, 1890, p. 386.
  A short description of the filters at St. Petersburg, and a comparison with those at Warsaw.
STURMHÖFEL. Zeitschrift f. Bauwesen, 1880, p. 34.
  A description of the Magdeburg filters, with plans.
TOMLINSON. American Water-works Association, 1888.
  A paper on filters at Bombay and elsewhere.
TURNER. Proc. Inst. Civil Engineers, 1890, c. p. 285.
  Filters at Yokohama.
VAN DER TAK. Tijdschrift van de Maatschapping van Bouwkunde, 1875(?).
  A description (in Dutch) of the Rotterdam water-works, including the wooden drains which caused the trouble with crenothrix, and which have since been removed. Diagrams.
VAN IJSSELSTEYN. Tijdschrift van het Koninklijk Instituut van Ingenieurs, 1892–5, p. 173.
  A description of the new Rotterdam filters, with full drawings.
VEITMEYER. Verhandlungen d. polyt. Gesell. zu Berlin, April, 1880.
  Filtration and purification of water.
WOLFFHÜGEL. Arbeiten aus dem Kaiserliche Gesundheitsamt, 1886, p. 1.
  Examinations of Berlin water for 1884–5, with remarks showing superior bacterial efficiency with open filters.
——— Journal für Gas- u. Wasserversorgung, 1890, p. 516.
  On the bacterial efficiency of the Berlin filters, with diagrams.
ZOBEL. Zeitschrift des Vereins deutsche Ingenieure, 1884, p. 537.
  Description of filters at Stuttgart.

OTHER LITERATURE.

Many scientific and engineering journals publish from time to time short articles or notices on filtration which are not included in the above list. Among such journals none gives more attention to filtration than the *Journal für Gasbeleuchtung und Wasserversorgung*, which publishes regularly reports upon the operation of many German filters, and gives short notices of new construction. The first articles upon filtration in this journal were a series of descriptions of German water-works in 1870-73, including descriptions of filters at Altona, Brunswick, Lübeck, etc. Stenographic reports of many scientific meetings have been published, particularly since 1890, and since 1892 there has been much discussion in regard to the "Rules for Filtration" given in Appendix I.

A Report of a Royal Commission to inquire into the water-supply of the metropolis, with minutes of evidence, appendices, and maps (London, 1893-4), contains much valuable material in regard to filtration.

The monthly reports of the water examiner, and other papers published by the Local Government Board, London, are often of interest.

The German "Verein von Gas- u. Wasserfachmänner" prints without publishing a most useful annual summary of German water-works statistics for distribution to members. Many of the statistics given in this volume are from this source.

Description of the filters at Worms was given in the *Deutsche Bauzeitung*, 1892, p. 508; of the filters at Liverpool in *Engineering*, 1889, p. 152, and 1892, p. 739. The latter journal also has given a number of descriptions of filters built in various parts of the world by English engineers, but, excepting the articles mentioned in the above list, the descriptions are not given in detail.

INDEX.

	PAGES
Altona, freedom of, from cholera	148
water-supply of	146, 171
Alum, use of in filtration	88, 109
objections to the use of	110
Alumina, precipitated, use of	88, 113
American cities, water-supplies of, and typhoid fever in	126
Amsterdam, filters at	178
Anderson process	114, 178
Antwerp, filters at	178
use of alum at	110
Area of filters to be provided	43, 142
Bacteria, apparent and actual removal of, by filters	83
from underdrains	83
in Elbe at Altona	172
in fæces	133
in water	80, 134
method of determining, in water	140
number to be allowed in filtered water	140
of cholera in river water	149
of typhoid fever, life of, in water	134
of special kinds to test efficiency of filtration	82
to be determined daily	140
Bacterial efficiency of mechanical filters	107, 111
examination of water	89, 140
Berlin, apparatus for regulating depth of water on filters	55
cholera infantum due to imperfectly filtered water	147
friction in underdrains of filters	40
regulation of rate of filtration	49, 51
water-works	167
Boston, protection of purity of water-supply	104
water-works, experimental filters	69
Bottoms of filters must be water-tight	143
Breslau, filters at	180
Brussels, ground-water supply of	182
Budapest, filters at	180
Burton, regulation of rate at Tokio, Japan	54

# INDEX.

	PAGES
Chemnitz, intermittent filtration at	100
Chicago, reduced death-rate with new water intake	135
Cholera infantum due to impure water	144, 173
Cholera, in Hamburg due to water	148
removal of danger from	122
Clark, H. W., sand analyses made by	22
Clark's process for softening water	88
Cleaning filters	64
Cologne, water-supply of, from wells	182
Coloring matter, removal of, by filtration	88
Continuous filters	5
filtration, nature of process	88
Cost of filters and filtration	4, 16, 118, 176
effect of rate upon	44, 121
Cost of Lawrence filter	98
Cost of operation of filters	121
Covered filters, efficiency of	17
Covers for filters	12, 15
needed at Altona	147
in the United States	17
omitted at Lawrence	97
Crenothrix	170, 179
Diarrhœa due to impure water	144
Double filtration at Schiedam	179
Dresden, water-supply of, from filter-gallery	182
Effective size of sand	20, 156
European sands	23
Efficiency of filtration	79, 84, 87
effect of rate upon	46
effect of size of sand-grain upon	25
effect of thickness of sand-layer upon	30
European filters	87
Effluents, wasting after scraping	70, 141
Elbe, watershed of	171
European filters, cost of	118
Fæces, number of bacteria in	133
Ferric salts, use of	113
Filling sand with water from below	64
Filter-beds, bottoms of, must be water-tight	12, 143
covers for	12
form of	11
size of	10
Filtered water, number of people supplied with	3
Filtering materials	19
Filters at Altona	172
at Berlin	169
at Hamburg	176
at London	164

INDEX.   193

	PAGES
Filters first constructed at London	79
for household use	115
general arrangement of	6
regulation of rate of filtration of	48, 142
statistics of, at various cities	159
Filtration, degree of purification required	5
general nature of	5
FitzGerald, Desmond	69, 104
Fränkel and Piefke, experiments on removal of disease-germs	82
Frankfort on Main, water-supply of, from springs	182
Frankland, Dr. Percy	80
Friction of filtered water in pipes	170
water in gravel	33
water in sand	21
Frictional resistance of underdrains	36
Frost, effect of, at Altona	13, 147, 172
effect of, upon filters	12, 147
Frühling, on the heating of water by sunshine	16
underdraining at Königsberg	35
Fuller, G. W.	133
German Imperial Board of Health	30, 47, 50, 71, 91
regulations in regard to filtration	139
Glasgow, water-supply of, from Loch Katrine	181
Gravel, layers	31
friction of water in	33
screening of, for filters	33
Ground-water supplies	3
the use of, in Europe	182
Halbertsma, H. P. N.	50, 55
Hamburg, apparatus for regulating depth of water on filters	55
health of	144
regulation of rate of filtration	52
underdrains of filters at	38
water-supply of	2, 146, 175
Hardness, removal of	88
Havel, watershed of river, above Berlin	168
Household filters	115
Ice on filters	13
Impounding reservoirs	2, 131
Intermittent filtration	93
application of	104
at Chemnitz	100
at Lawrence	96
of Pegan Brook	104
Iron, metallic, the use of	114
Kirkwood, James P.	8, 32, 43, 47, 51, 57, 59, 63
Koch, Dr. Robert	147, 149
Kraus, Dr., on the cause of cholera infantum	146

INDEX.

	PAGES
Kümmel	46, 47, 82
Lawrence City filter, description of	96
Lawrence Experiment Station	93
air in water filtered in winter at	42
depth of sand removed at	66
depth of water on filters	42
effect of loss of head upon efficiency	57
effect of size of sand-grain upon efficiency	28
effect of size of sand-grain upon frequency of scraping	28
efficiency of filters at various rates	46
efficiency of filtration at	82, 85
experiments with continuous filtration	103
filters of fine sand	27
filters with various sand-grain sizes	28
gravel for filters at	35
growth of bacteria in sterilized sand at	81
intermittent filtration investigated	93
method of sand analysis at	19, 151
quantities of water filtered at various losses of head	62
wasting effluents not necessary	71
Lawrence, typhoid fever at	98, 129
Lea, watershed of river, above London	163
Leipzig, water-supply of, from wells	182
Lindley	39, 47, 50, 53, 77
Literature of filtration	183
Loam in filters	31
London, water-supply of	79, 161
sewage treatment on the **watersheds of the Thames and Lea**	9, 162, 163
Loss of head	48
limit to	56, 63, 142
reasons for allowing a high	61
Magdeburg filters at	179
Manchester, water-supply **of**	181
Massachusetts State Board of Health (*see* Lawrence Experiment Station).	
Mechanical filters	106, 137
reasons for the use of	108
Mechanical filtration with alum	109, 137
Metallic iron, use **of**	114
Mills, H. F.	93, 95, 98
Müggel (Berlin), filters **at**	55, 167
Munich, water-supply of, from springs	182
Nitrification, effect of, upon bacteria	94
Objects of filtration	122
Oder, watershed of, above Breslau	180
Odors, removal of, by filtration	122
Organic matters in water	79
Paper manufacturing, filtration of water for	5, 106
Paris, ground-water **supply of**	182

	PAGES
Passages through the sand in filters.................	63
Pegan Brook, purification of.......................	104
Piefke............................44, 46, 50, 59, 65, 69, 70, 71, 76, 80, 81, 86
Plügge and Proskower..............................	80
Plymouth, Penna., typhoid-fever epidemic at.........	132
Pollution of European water-supplies.................	89
Polluted waters, utilization of excessively............	103
Porcelain filters for household use...................	116
Quantity of water per capita in America..............	119
Rate of filtration.............................43,	142
at various places........................	160
effect of, upon cost..................44,	121
effect of, upon efficiency..................	46
lower after scraping.....................	72
regulation of...........................	48
Regulation of filters..........................48,	142
old forms of regulators....................	48
modern forms of regulators................	50
Reincke, Dr., report on health of Hamburg for 1892...	144
Renisch on the cause of poor filtration at Altona......	173
Reynolds, Dr. A. R., on Chicago's water-supply.......	135
River waters, the use of............................	130
Roofs for filters...................................	16
Rotterdam, filters at...............................	178
St. Louis, regulators for proposed filters..............	51
St. Petersburg, filters at...........................	181
Samuelson..	47
Sand..	19
analyses of European...........................	23
analyses of, from leading works.................	26
appliances for moving..........................	64
compactness of, in natural banks.................	57
depth of, in filters.........................30,	81
depth to be removed from filters.................	65
dune...	22
dune, washing of, impossible....................	78
effect of grain size upon the frequency of scraping.	28
effect of grain size upon the efficiency of filtration.	25
effective size of........................... 20,	156
extra scraping before replacing fresh..............	67
for filtration............................19,	29
friction of water in............................	21
grain size of..............................19,	151
in European filters.............................	22
in Lawrence filters, two sizes of..................	96
method of analysis of...........................	151
quantity to be removed by scraping...............	70
replacing.....................................	67

INDEX.

	PAGES
Sand selection of	29
size of passages between grains of	6
thickness of layer	30, 143, 173
uniformity coefficient	20
Sand washing	19, 72
cost of	77
drum washers	74
ejector washers	75
Greenway's machine	75
hose washing	73
Pegg's machine	74
water for	76
Sandstone filters for household use	116
Schiedam, double filtration at	179
Scraping filters	7
amount of labor required for	77
frequency of	45, 68
frequency of, at various places	160
Sedgwick, W. T.	82
Sediment, removal of	88
Sediment layer	6, 27
influence of on bacterial purification	80
thickness of	29, 62
Sedimentation basins	8, 164, 175
effect of	137
Sewage treatment above London	9, 162, 163
Simpson, James	79
Spree, watershed of, above Berlin	167
Statistics of some filters	159
Storage, effect of, upon quality of water	136
Storage reservoirs, water from	2, 131
Stralau (Berlin), filters at	49, 167
Surface-water, the use of unfiltered, in Europe	181
Tastes, removal of	122
Tegel (Berlin), filters at	51, 167
Thames, watershed of, above London	161
Theory of continuous filtration	79
Tokio, regulation of rate at	54
Trenched bottoms for filters	32, 36, 96
Turbidity, removal of	88, 122
Typhoid fever, at Berlin and Altona	12, 81, 135, 172
carried by personal contact	135
from the use of water filtered mechanically	108
in American cities	126
in Chicago	135
in Lawrence	98
in London	165
loss from	124

                                                                                    PAGES
Typhoid fever, prevention of.. 123
            reduction of, by filtration at Zürich........................ 181
Typhoid-fever germs, life of, in water.. 134
Underdrains..31, 35
            friction of, in Lawrence filter................................. 96
            size of.. 37
            ventilators for.. 40
Uniformity coefficient of sand..20, 156
Ventilators for underdrains.. 40
Vienna, water-supply of, from springs.. 182
Warsaw, filters at... 181
            friction in underdrains.. 39
            regulation of rate at.. 53
Wasting effluents..70, 141
Water, depth of, upon filters.. 41
            heating of, in filters... 41
            organic matters in... 79
Water-supplies of American cities.. 126
Water-supply and disease... 133
Waters, what require filtration.. 130
Winter, effect of, upon filtration.....................................12, 146, 172
            intermittent filtration in................................102, 104
            temperatures at places having open and covered filters......... 15
Zürich, filters at... 180

www.ingramcontent.com/pod-product-compliance
Lightning Source LLC
Chambersburg PA
CBHW020858230426
43666CB00008B/1231